# Unlocking Energy Innovation

# Unlocking Energy Innovation

## How America Can Build a Low-Cost, Low-Carbon Energy System

Richard K. Lester and David M. Hart

The MIT Press
Cambridge, Massachusetts
London, England

MIT Press books may be purchased at special quantity discounts for business or sales promotional use. For information, please email special_sales@mitpress.mit.edu or write to Special Sales Department, The MIT Press, 55 Hayward Street, Cambridge, MA 02142.

This book was set in Sabon by the MIT Press. Printed and bound in the United States of America.

Library of Congress Cataloging-in-Publication Data
Lester, Richard K. (Richard Keith), 1954–
Unlocking energy innovation : how America can build a low-cost, low-carbon energy system / Richard K. Lester and David M. Hart.
p. cm.
Includes bibliographical references and index.
ISBN 978-0-262-01677-3 (hardcover : alk. paper)
1. Energy industries—Technological innovations—United States. 2. Renewable energy sources—Technological innovations—United States. 3. Technology and state—United States. I. Hart, David M., 1961– II. Title.
HD9502.U52L475 2012
333.790973—dc23

2011031297

10 9 8 7 6 5 4 3 2 1

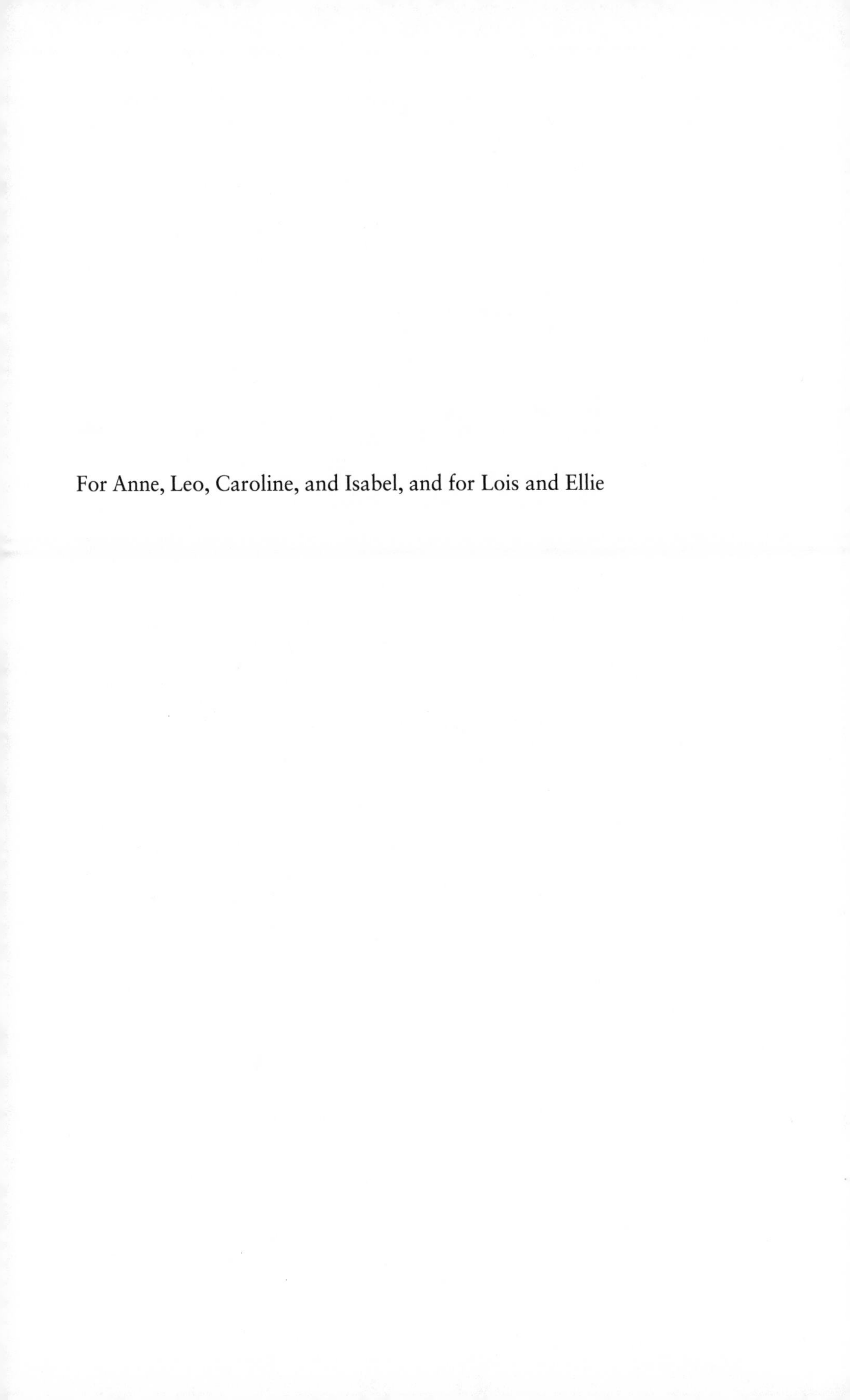

For Anne, Leo, Caroline, and Isabel, and for Lois and Ellie

# Contents

# Preface

The last few years were not good ones for those seeking to accelerate the transition to a low-carbon energy system. Ambitious plans to regulate greenhouse gas emissions on a global basis fizzled. Negotiations in Copenhagen in 2009 to frame a successor to the Kyoto Protocol produced a fig-leaf agreement that failed to conceal profound divisions among the major countries. A subsequent climate conference in Cancun accomplished little.

Similarly ambitious plans within the United States stalled, too. The Waxman–Markey bill, which sought to commit the United States to deep cuts in emissions, died in the Senate after a bitter passage through the House of Representatives. The Obama administration's 2009 stimulus package, which included unprecedented investments in energy efficiency and new energy technologies, was largely spent by mid-2011. As fiscal pressures mounted there seemed to be little prospect that follow-on funds would be found to continue these efforts.

Three shocks to the global energy system diverted attention away from the climate change agenda as well. The largest offshore oil spill in U.S. history dominated the headlines through the summer of 2010. Less than a year later, in March 2011, the destruction of four reactors at Japan's Fukushima Nuclear Power Station dealt a heavy blow to hopes that nuclear energy could provide a near-term alternative to fossil fuels for electricity generation. And the riveting spectacle of Arab revolution and upheaval burst onto the world stage in January 2011 and swept across the Middle East and North Africa. With U.S. forces involved in hostilities in Libya and gasoline prices approaching four dollars a gallon at the time of this writing, the focus of U.S. policy has shifted back to energy security and the goal of reducing oil imports.

Neither the failure of global and U.S. climate policy nor the emergence of other pressing issues changes the facts about climate and energy. Business as usual is unsustainable over the long run. The build-up of greenhouse gases in the atmosphere continues unabated, and its environmental and human costs are mounting. For each year of delay it will be necessary to "bend the curve" of carbon emissions that much more sharply if the worst consequences of climate change are to be averted.

Yet recent events also reveal an important truth that advocates of the need for a low-carbon energy transition must face. Broad public support for bold action does not exist in the United States. Until it does, no long-term, comprehensive national policy will be sustainable. The energy transition will require the exertion of public authority on a large scale, and it will touch the lives of almost everyone in a direct and tangible way. If the consent of the governed is granted only grudgingly or not at all, such a program will inevitably be hamstrung by resistance and will remain at constant risk of reversal.

The energy/climate debate of recent years has been dominated by the idea that the price of fossil fuels must be raised to reflect the full cost of their use. What is needed now is a different focus for the debate, one with the potential to build public support over time. *Innovation* should become that focus. A carbon price must surely be a part of any energy transition strategy. But the pain imposed by the stick of higher prices must be counter-balanced—in the public mind and in practice—by the carrot of better, safer, and (ultimately) cheaper energy services made possible by new technologies, new business models, and new institutions.

There is no escaping the fact that remaking a significant fraction of the economy—which is what "bending the curve" of carbon emissions will entail—will inflict real costs and that firms, industries, workers, and communities which stand to be adversely affected will mount vigorous opposition. But if attention can be focused on the opportunities presented by the energy transition, history provides some reason for hope. Over the last century, the United States built large, stable innovation systems that drove extraordinary technological transformations in comparably vast regions of the economy, including agriculture, defense, health, and telecommunications. Enlightened public policy, sustained by durable political coalitions, enabled each of these transformations. Energy must now join this list.

That may seem like an unlikely prospect. But in the field of energy there are always surprises. We are mindful that what now seems inconceivable may be tomorrow's conventional wisdom, and that some of today's seemingly unalterable political realities will one day be quaint artifacts. The energy transition that is our focus in this book will take decades to complete, and although we cannot ignore immediate crises and urgent political imperatives, our task is also to look beyond them.

# Acknowledgments

A large number of people have contributed, directly and indirectly, to this book. It is based on the findings of the Energy Innovation Project, a three-year study by a research team based at the MIT Industrial Performance Center (IPC). We are indebted to the Doris Duke Charitable Foundation for its generous financial support of the project.

We were fortunate to be guided by an outstanding advisory board, chaired by Charles Vest and also including Ralph Cavanagh, Ray Lane, Bill Madia, Paul Maeder, John Podesta, John Reed, Jim Rogers, Steve Specker, Rob Stavins, John Sununu, and Mason Willrich. None of these advisors agreed with all of our conclusions, and some disagreed strongly with some of them, but all were generous with their wisdom and their time, and we are tremendously grateful to them.

At MIT our colleagues John Deutch, Ernie Moniz, and Steven Ansolabehere (now at Harvard) served as an internal advisory council, and their encouragement, incisive critiques, and constructive suggestions were gratefully received.

The Energy Innovation Project research team at the Industrial Performance Center included Howard Herzog, Harvey Michaels, Michael Piore, Elisabeth Reynolds, Rohit Sakhuja, Ed Steinfeld, and Erika Wagner and, at Harvard Business School, Shikhar Ghosh and Ramana Nanda. Rebecca Henderson and Richard Newell (before departing for government service) led an affiliated project under the auspices of the National Bureau of Economic Research. Mason Willrich, in addition to serving as a senior advisor to the project, also participated actively in the research. We are indebted to all of these colleagues for their valuable contributions to the study.

Our talented students at MIT and George Mason University included Kathy Araujo, Phech Colatat, Ed Cunningham, Kat Donnelly, Ashley Finan, Michael Hamilton, Kadri Kallas, Valerie Karplus, Sherif Kassatly, Kevin Knight, Florian Metzler, Jonas Nahm, Will Norris, Neil Peretz, Anil Rachakonda, Alice Rosenberg, Genevieve Borg Soule, Rachel Wellhausen, and Yuan Xu. We thank them all for their skill and hard work.

A large group of experts from industry, government, and academia participated with us in a stimulating series of project workshops—on energy efficiency in commercial buildings and in residential buildings, on carbon capture and sequestration, on the smart grid, on ARPA-E, and on the development of the New England clean energy cluster. These experts, too numerous to mention by name here, gave freely of their time and expertise, and we are grateful to them.

Over the course of the project we benefited greatly from discussions with many colleagues who were generous with their insights and suggestions, including Daron Acemoglu, Bill Bonvillian, Gaston Caperton, Armond Cohen, Phil Deutch, Jeff Eckel, Dan Goldman, David Goldston, Bill Helman, Christopher Hill, Marija Ilic, Steve Isakowitz, Ralph Izzo, Paul Joskow, Kent Larson, Richard Nelson, Ron O'Hanley, John Parsons, Ignacio Perez-Arriaga, Arati Prahkakar, Burt Richter, Ray Rothrock, Maxine Savitz, Dick Schmalensee, Bob Solow, Joel Spira, Hemant Taneja, Mitch Tyson, and Jim Wolf.

Marika Tatsutani provided valuable editorial advice during the later stages of the project.

Anita Kafka as usual found countless ways to help move the project forward while keeping the Industrial Performance Center running smoothly. Carol Sardo ably assisted during the project's later stages.

At MIT Press we are indebted to John Covell and Ellen Faran and the staff of the Press for the care and attention they have given to this book.

# Introduction: Zero to Eighty in Forty

America needs energy. We need energy for many reasons—to stay warm, to stay cool, to work, to get around, to communicate, and for much, much more. We need energy to be reliable and readily available. We need its price to be low and stable, so businesses can stay competitive and plan ahead, and so families don't have to worry about utility bills eating up the household budget.

To meet these many needs, our nation has built a huge and intricate system for producing and delivering energy on demand in a variety of forms. The system has worked reasonably well from the perspective of energy users. To be sure, there have been periods of high and volatile prices, as well as occasional long lines at the gas pumps and power blackouts. These episodes have been disruptive, but they have also been exceptions to the rule. By and large energy in America has been available, reliable, and cheap.

Viewed from a broader perspective, however, the American energy system has worked less well. Extraction of energy resources has provided many good jobs, but it has also inflicted hardships on workers, communities, and the environment. Energy consumption has yielded pollution and waste, as well as needed energy services. Securing the nation's overseas energy supply lines has been a major factor in U.S. foreign and military policy, and it has been implicated in two of the nation's recent wars. Spending on oil imports has helped to prop up unfriendly governments and is the single largest contributor to the U.S. trade deficit.

In the past these costs have been judged to be tolerable, an acceptable exchange for the many benefits of available, reliable, and cheap energy. But that judgment is no longer tenable. Chief among the many energy-related challenges that we face is climate change, which results primarily from the burning of fossil fuels. These fuels today account for more than 80 percent of the world's energy supply. If "business as usual" continues—if today's trends are extrapolated into the future and nothing is done to control carbon dioxide and related greenhouse gas emissions that are by-products of fossil fuel use—before the end of this century the world is likely to experience climate change on a devastating scale.

Climate change has recently become an unfashionable subject in Washington. Elected officials who are prepared to deny categorically that the problem exists have grown more numerous. Others would not go so far, but nevertheless assign greater weight to other problems, such as dependence on imported oil or the energy system's local and regional environmental impacts. These are serious problems, but controlling the atmospheric build-up of carbon dioxide is the most serious of all. It is also the most difficult energy problem to solve and the most urgent one to address. At a minimum, we believe that an aggressive effort to reduce carbon dioxide emissions would not worsen any other energy-related problem and would have collateral benefits with respect to many. A low-carbon energy system would necessarily mean drastically reduced oil imports, for example.

To avoid the most harmful effects of rising greenhouse gas concentrations while still meeting the growing demand for affordable and reliable energy services, nothing less than a fundamental transformation of current patterns of energy production, delivery, and use on a global scale will be required. The best available science tells us that to avert the worst consequences of climate change, greenhouse gas emissions will need to be cut significantly by mid-century. To reach this goal, the delivery and use of energy must become much more efficient, and fossil fuels must be replaced by low-carbon energy sources in most applications. If this vast effort is to succeed, the United States, the world's largest economy, second largest user of energy, and greatest source of new ideas and new ways of doing things, must lead the way. President Obama has proposed that we cut carbon dioxide emissions by 80 percent in the next 40 years (hence the title of this chapter: "Zero to Eighty in Forty").

It will be very hard to achieve this goal. Most of the technologies available today to support the energy transition are too expensive, are too difficult to scale, or have other environmental or economic drawbacks. Even when cost-effective technologies are available, their deployment is often blocked by organizational or institutional barriers. Only through innovation can these problems be overcome. That means innovation in many spheres: in technology, in business models, in institutions, and in government policy.

## Three Waves of Innovation and an Innovation System to Drive Them

The energy system will not be transformed all at once, nor will it be transformed by a single "magic bullet" solution. Instead, we envision the energy transition in the United States unfolding in three waves. Each wave will gather momentum at a different rate, and each will sweep over the energy sector at a different time. But they must all be accelerated and pursued in parallel, and in all cases work must begin right away:

1. The first wave, ramping up in this decade and continuing beyond, must focus mainly on energy efficiency gains, especially in buildings, which currently account for about 40 percent of total energy use and 70 percent of electricity use. Although some additional technological advances may be needed, many options for more efficient heating, lighting, air-conditioning, insulation, and other energy uses are already available; therefore, the primary innovations in this wave will be institutional and organizational.

2. The second wave will overlap with the first, but will have its largest impact between 2020 and 2050. It must focus on the large-scale deployment of known low-carbon technologies for electricity generation, transmission, distribution, and end-use, such as nuclear, solar and wind power, carbon capture and sequestration (CCS), electric/hybrid transportation, and grid-scale storage, driving down their costs through continual innovation.

3. A possible third wave of innovation, achieving scale only in the second half of the century, may result from radical technical advances generated by fundamental research in a broad range of scientific fields. This research must be generously funded from now on.

America does not yet have an innovation system in place today that is capable of driving forward any of these waves. By "innovation system," we mean a set of institutions—organizations, rules, policies, relationships—that generates and diffuses new ways of doing things. The U.S. innovation system as a whole has been extraordinarily productive in other fields, such as information technology and biotechnology. Even in energy, the historical record of innovation over the long term, going back to Edison and before, is impressive. But in the past few decades, energy innovation has been lagging in this country. The firms and financial institutions that lie at the core of the U.S. innovation system in other fields are risk-averse in the energy field, with a history of underinvestment in research and development and often a strong interest in preserving the status quo. Public institutions, which play a critical supporting role in the U.S. innovation system, have not tackled energy issues with the seriousness and the resources needed to trigger fundamental change.

Our goal in this book is to set out a new institutional architecture for America's energy innovation system. The public sector must be the primary instigator of institutional change. Markets left to their own devices will not take climate change into account. But the purpose of public action ought not to be the creation of a system in which government agencies are the central actors. Rather, its purpose should be to unlock the immense skills and resources of America's private entrepreneurs, investors, producers, and energy users, so that they carry forward most of the innovation tasks. Unlocking: that is the operative verb and thus the metaphor that provides our title.

## The Plan of the Book

Our approach departs from the most familiar positions on this subject. We set the stage for it in chapter 1, where we provide a framework for thinking about the scale of the problem. The facts are daunting, and we must not rely on wishful thinking. For the United States to go from "zero to eighty in forty," the country will have to dramatically accelerate innovation to improve energy efficiency *and* to replace fossil fuels with very low-carbon energy sources. The facts lead to an innovation agenda focusing on the electricity sector, which will play an even more central role in a decarbonized energy system than it does now. Neither recent advances in the natural gas industry nor the expansion of first- and second-generation biofuels will serve as more than a bridge to this electricity-centric system over the next few decades. We argue that the country must begin to tackle the energy innovation agenda now, because the pace of change in such a large, vital, and complex system is inevitably slow. Forty years is not that long a time to test, refine, and get innovations into practice on a mass basis. Breakthroughs that are still in the laboratory today will not make a major difference on this timescale. The real energy innovation opportunity lies in improving technologies whose basic scientific and engineering characteristics are already well known and embedding them in supportive economic, social, and organizational settings.

Chapter 2 is about how energy innovation occurs. We describe a stylized four-stage process of innovation in which many different kinds of institutions, including firms, investors, research organizations, government agencies at all levels, and users of diverse sorts, participate. We map this stylized description onto the system as it has operated in the United States in recent years and find a number of gaps and problems, especially in the demonstration and early deployment stages. We then assess the conventional wisdom about what to do. The Manhattan Project (which built the first atomic bomb) and the Apollo Program (which put Americans on the moon) are too government-centric to serve as adequate models for revamping the U.S. energy innovation system, as some enthusiasts of these successes of the past have recommended. We have more sympathy for the position that a price be put on carbon emissions, either through a tax or a cap-and-trade system, but we see such a price as necessary rather than sufficient to stimulate innovation on the scale and in the direction required. We also accept the argument that public funding for R&D should be expanded, but that too is far from sufficient, given the complexity of the energy innovation challenge and the many roles that R&D ought to play.

This review of what others have proposed leads into our own proposals. Building on a set of principles articulated at the end of chapter 2, chapters 3 to 7 set forth specific institutional arrangements that we believe can drive all three waves of innovation. Chapter 3 makes the case for restructuring the electric utility sector (a central player in all three waves), so that new entrants with new business models and technologies flow into it. Following the logic that has swept through industry after industry over the past thirty years, the restructuring process would create competition within layers of the electricity system that have long been sheltered from it, such as power generation and end-user energy management. The footprint of the electric utility would shrink markedly, and its central task would shift to being a "smart integrator" of low-carbon supply and demand resources that plug into the grid. The unique features of the electricity system mean that such a restructuring would have to be carefully designed and continually overseen. Regulators, like utilities, must get smarter in the twenty-first century.

In chapter 4, we turn to the first wave of innovation. Rapid improvements in energy efficiency in the economy as a whole depend especially on large improvements in building energy efficiency. Such improvements will require regulatory innovation. We argue for the enactment of energy codes for new buildings across all the states, supported by innovations in technology and design. Appliance energy efficiency standards, which are a federal responsibility, must also be ratcheted up, a process the Obama administration has restarted after a long period of inactivity. Retrofitting of existing buildings presents the most difficult element of the building efficiency puzzle. Innovations in financing that attract more private capital into the market along with new ways of servicing loans and better information offer the best opportunities to deal with it.

Chapters 5 and 6 turn to the second wave of innovation. The core of chapter 5 is a novel regionally based structure for demonstrating and driving down the costs of low-carbon energy supply technologies, the key objective of the second wave. It would comprise three new institutions. A network of funding authorities ("Regional Innovation Investment Boards") would decide which large-scale demonstration projects and early adoption programs to support. Their members would mainly be downstream users of energy innovations in each region. A federal "gatekeeper" agency would certify the proposals that could be placed before the regional boards. It would ensure that the innovations that the boards target have the potential to meet the goals of carbon dioxide emissions reduction, affordability, and reliability. A surcharge on electricity sales would fund these investments. Rather than having an entitlement to the proceeds of the surcharge, the regional boards would have to compete to win support for their portfolios from state-level "trustee" organizations. This three-part structure would introduce multiple levels of competition into a process that is now laden with bureaucratic and political baggage. It would also insulate the funding from the vagaries of federal budgeting.

Chapter 6 considers how to advance innovation in distributed power generation, storage, and the "smart grid." These innovations have the potential not only to enhance the reliability and resilience of the electric power system, but also to engage small-scale users in energy innovation in ways that have not been possible before. We support an open architecture for the grid, one that gives customers along with small-scale power generators and storage providers more control than they have now. Customers will be incentivized to exercise that control by prices that vary over time, depending on supply and demand conditions in the electricity market. Most of the costs of this kind of innovation will be repaid by savings on electricity bills or by distributed sales into the grid, although the regional boards may also play a supporting role. The chapter also looks briefly at the utility side of the smart grid, which involves (among other things) hardware and software to manage the new power flows and fluctuations stimulated by the customer-controlled architecture. The regional boards might play a valuable role here, too.

Chapter 7 addresses the third wave, a set of innovations that will play a big role after 2050. While no one knows what these will be, everyone should know that there will be a need for continued energy innovation on a long time scale. The country, and the world, will not find these "game-changers" without steady investment. We suggest a pluralistic and open system of public research funding that encourages international collaboration. There is an important place for potential users in the creation of long-term options as well so that this part of the innovation system is informed by downstream requirements. Chapter 8 concludes the book, highlighting our most important recommendations and the logic that ties them together.

Although we discuss many of the new technologies and business models that are likely to play a role in the energy transition, this book is not about which specific kinds of devices or companies should receive private investment or government support. Nor is it about whether the government or the market is better at stimulating and choosing innovations. The energy innovation system necessarily involves both, as well as hybrid institutions that combine public and private action. The question we try to answer in this book is which particular combinations of institutional arrangements are likely to be most effective in driving energy innovation in the directions that it needs to go over a period of decades.

The book focuses on American institutions and has little to say about the innovation system at the global level. The domestic focus is not meant to diminish the importance of activities and policies elsewhere in the world. Many energy markets are global, the technical and business communities involved in the energy industry are global, innovation networks are becoming increasingly global, and climate change is the ultimate global problem. The United States has a tremendous stake in the success of low-carbon energy innovation in countries like China and India, and American participation in those efforts will be to mutual benefit. But each country's innovation system is unique, shaped by the particularities of its history, economy, and politics. That is certainly true of the United States. Although we strongly endorse learning from abroad and closer international cooperation, American institutions will change mainly in response to domestic influences and along pathways that reflect this country's special characteristics.

Whatever happens elsewhere, U.S. leadership in energy innovation will be essential to the success of the world's climate change mitigation efforts. International cooperation is a complement, not a substitute, for American creativity, resourcefulness, and entrepreneurship. This book is about how to mobilize America's enormous innovation resources in the service of a decades-long, global energy transition. The book is about a long game, and it is particularly about that part of the long game that will be played here at home.

In order to lay the groundwork for all of this, we need first to describe what is known about climate change and energy innovation. That is the subject of chapter 1, to which we now turn.

# 1

# Beyond Wishful Thinking: Facts, Deductions, and Grounded Assertions About Climate and Energy

The late Senator Daniel Patrick Moynihan is reputed to have said that "everyone is entitled to his own opinions but not his own facts." The debate over climate and energy is muddled by confusion about the facts. Some confusion has been sown deliberately by actors whose interests are at odds with the facts. Some is the inevitable result of the extraordinary scale and complexity of the natural and social systems at issue. This chapter sets forth the facts as we understand them. Facts are statements about the present and the past that have been validated by evidence. We include in this category the impact of greenhouse gas emissions from fossil fuel combustion on the global climate in the recent past and the ubiquity and immense scale of fossil fuel use in the United States and elsewhere in the industrialized world.

There are no facts about the future, of course. But we can use mathematics and logic to make deductions about the future from the facts. For instance, we can deduce that global carbon dioxide emissions reductions of greater than 80 percent will be impossible if the United States does not reduce its emissions since it is a fact that the U.S.'s share of global emissions is 20 percent. This chapter engages in this sort of deduction as well. We deduce that there are limits to the pace of change in a system that is as big, important, and complicated as our energy system. We deduce that any national effort to achieve the goal of "80 in 40" (an 80 percent reduction in carbon dioxide emissions in 40 years) will test these limits, both in improving energy efficiency and in expanding low-carbon energy supplies.

We also make assertions in this chapter that are grounded in facts and logic but involve judgments as well, based on research carried out by us and others. We assert that pushing for "80 in 40" is a good idea, even though it will be hard to achieve this goal, because not pushing for it will likely lead to bad outcomes of one sort or another. We assert that this goal is likely to be compatible with continued affordability and reliability of energy in the United States and that it can be achieved without undermining other long-standing goals of energy policy—indeed, that it will likely advance at least some of those goals. We assert that neither natural gas nor biofuels nor some yet-to-be discovered breakthrough provides a plausible pathway to achieving "80 in 40." We assert that the pursuit of this goal will accelerate the electrification of the U.S. energy system. And we assert, in conclusion, that an "80 in 40" energy innovation strategy should emphasize cutting the cost, enhancing the reliability, and speeding the uptake of low-carbon technologies whose basic scientific and engineering characteristics are already well known, especially technologies in the electricity sector.

Readers who accept these facts, deductions, and grounded assertions may want to skip ahead to chapter 2. Such readers will undoubtedly find much to argue with there and in later chapters. Readers who are skeptical, need some convincing, or have not thought through the problem before will find a fuller exposition in the remaining sections of this chapter.

## The Climate and Energy Challenge: Immense Scale, Enormous Scope

Much about the past and future of the earth's climate is still uncertain, but several facts are incontrovertible. The earth is warmer than it has been for a long time. The average global surface temperature has been higher over the last few decades than during any comparable period over the last 400 years and perhaps the last millennium.[1] All 20 of the warmest years since the instrumental record began in 1860 have occurred since 1981, and all 10 of the warmest years have occurred since 1998. In 2010, which tied with 2005 as the warmest year on record, the combined global land and ocean surface temperature was 0.62°C above the 20th century average of 13.9°C.[2] (See figure 1.1.)

The level of carbon dioxide in the atmosphere, meanwhile, has been rising steadily since scientists began taking measurements in 1957. There is no doubt that this increase has been caused mainly by the combustion of fossil fuels, which began in earnest at the start of the Industrial Revolution in the early nineteenth century. The total rise in the carbon dioxide

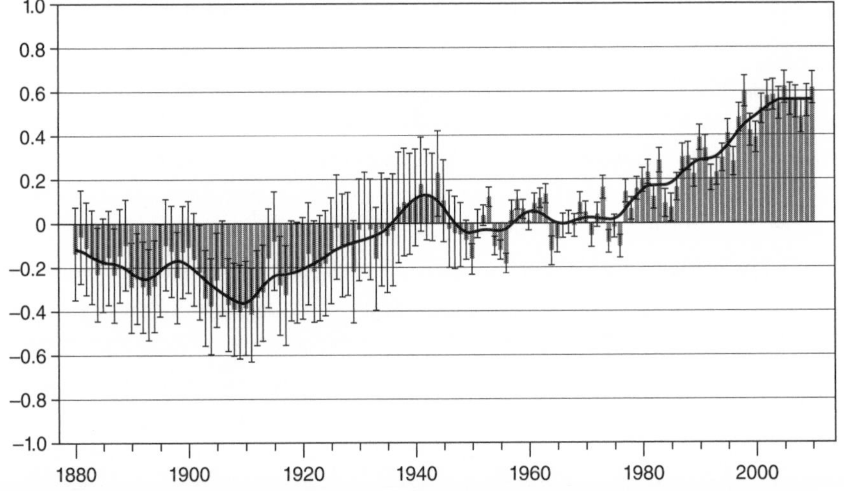

**Figure 1.1**

Global mean temperature over land and ocean surfaces, 1880–2010. *Source:* National Oceanic and Atmospheric Administration.

level over two centuries is about 40 percent, with most of that increase occurring in the last 50 years.

In 1965 the President's Science Advisory Committee (PSAC) recommended that the possible impacts of rising carbon dioxide on the global climate be studied. PSAC noted that carbon dioxide traps solar radiation, sending part of it back toward the earth's surface and causing net warming. (Later research showed that other substances also produce a similar effect, although their effects are in aggregate smaller than that of carbon dioxide.[3]) In recent years, a large international research effort has taken up PSAC's recommendation, investigating the relationship between the observed warming and the observed increase in atmospheric greenhouse gas concentrations. The conclusion of this effort is that it is highly likely (meaning that the probability is more than 90 percent) that most of the increase in the global average temperature since the mid-twentieth century is due to the increase in greenhouse gas concentrations, which in turn is mainly attributable to fossil fuel combustion.[4]

The rate of carbon dioxide emissions continues to increase globally. If one extrapolates from recent trends—the so-called "business-as-usual" scenario—the concentration of carbon dioxide in the atmosphere by 2100 will be roughly triple what it was before the Industrial Revolution began. How the earth's climate would respond to such an increase cannot be predicted with certainty, but the latest scientific estimates indicate that the global average temperature can be expected to rise by at least 4°C, and possibly by more than 6°C.[5] The high end of that range is roughly ten times the amount of warming observed so far and it is similar to the difference between today's climate and the coldest part of the last Ice Age, when ice sheets covered much of North America.

The projected consequences of climate change of this magnitude would be severe for both natural ecosystems and human societies. Scientists are already observing knock-on effects such as warming oceans, shrinking sea ice, more powerful storms, and the extinction of vulnerable species.[6] Another ninety years of climate change will likely put large coastal populations at greater risk of inundation due to rising sea levels and storms. (About 23 percent of the human population currently lives within 100 kilometers of the coast and less than 100 meters above sea level.) It will probably become harder for many people to grow their traditional crops, and their health will likely deteriorate because the range of pathogens will be extended. These burdens will fall most heavily on those least able to bear them.[7] To be sure, the great complexity of the earth's climate system means that such projections are uncertain. There is still a chance that the consequences of continuing down the business-as-usual path will turn out to be tolerable. But the weight of scientific evidence points in the other direction. Indeed, new findings suggest that the most widely accepted projections are underestimates.[8] In this sobering context, a combination of mitigation to reduce greenhouse gas emissions as aggressively as possible and adaptation to protect the most vulnerable human and natural systems may be the most sensible course. Or, as presidential science advisor John Holdren recently put it in a succinct summation, "We need enough mitigation to avoid the unmanageable, and enough adaptation to manage the unavoidable."[9]

Because of the many uncertainties surrounding the earth's climate system, as well as the divergent interests of different stakeholders around the world, there can be no exact answer to the question of what would constitute "avoiding the unmanageable" (or, put another way, what would be an acceptable upper limit on greenhouse gas concentrations in the atmosphere). Even so, many climate scientists have concluded that the worst risks of climate change might be avoided if the concentration of carbon dioxide could be kept below 550 parts per million (ppm), or roughly twice the preindustrial level. Although serious ecological and economic damage is likely even at this level, achieving it might avert truly catastrophic outcomes. If business-as-usual continues, 550 ppm will be reached by around 2050. At that point, preventing any further build-up of greenhouse gases in the atmosphere would require reducing emissions to zero overnight, since carbon dioxide stays up there for about a century on average. That of course would be impossible. Given the vast scale and slow rate of change of the world's energy system, we will need to begin to "bend the curve" toward lower carbon dioxide emissions well before the midcentury mark.

The longer the delay in getting started on curbing emissions, the steeper and more painful the cuts will have to be in later years. What happens during the next two to three decades, therefore, could well be decisive. If, by midcentury, the link between economic activity and carbon emissions has not been broken and significant progress toward decarbonizing the world's energy supplies still has not been made, the best scientific evidence indicates that we will have lost almost all chance of avoiding the very bad outcomes touched on above. The fact that the same evidence also holds out the possibility that it might not be as bad as all that, and that—if we are lucky—the consequences might be tolerable, makes the calculation a bit more complicated. But it does not change the basic conclusion that action on a timescale of a few decades will be needed to avoid a range of dangerous outcomes that at present seem much more likely than the more benign possibilities.

## Zero to Eighty in Forty: What It Would Take for the United States

U.S. participation is essential for the world to meet the energy-climate challenge. Although China recently surpassed the United States as the world's largest emitter of greenhouse gases, the United States still accounts for about 20 percent of total world emissions. And Americans remain among the world's largest emitters on a per capita basis.[10] It would be mathematically impossible for the world to stabilize carbon dioxide concentrations at twice the pre-industrial level without substantial cuts in U.S. emissions.

U.S. leadership would be much better than mere participation. It is highly improbable that the rest of the world will even try to transform the energy system unless the most powerful and richest country leads the way. U.S. leadership is not a sufficient condition for "avoiding the unmanageable and managing the unavoidable," but it is surely necessary. At every international climate change gathering, representatives from less-developed countries make the point that the United States and other industrialized economies have been responsible for most of the additions to the atmospheric inventory of carbon dioxide, and that their own future economic prospects will be unfairly curtailed unless the advanced economies shoulder more of the burden of future reductions in carbon dioxide emissions. Early in his presidency, President Obama joined with the leaders of the other G8 nations in pledging to cut emissions at least 80 percent by 2050. The G8 concluded that a cut of this magnitude would be needed to secure agreement on a global limit on carbon emissions. No global deal has yet materialized, but it seems clear that deep cuts in U.S. emissions will have to be part of one.

An effective strategy for reducing greenhouse gas emissions must focus on energy, which accounts for more than three quarters of U.S. emissions. The U.S. energy sector is vast, and the scale and scope of the energy transition will be comparably huge. Expenditures on energy use account for nearly 10 percent of the nation's GDP, and totaled $1.4 trillion in 2008.[11] About 85 percent of the energy that the United States uses today comes from fossil fuels. Almost all cars, trucks, and planes run on petroleum products. Coal, the most carbon-intensive fuel, is used to generate about half the country's electricity. The big energy supply sectors are among the most capital-intensive of industries, with large-scale production and processing facilities and massive pipeline, power grid, and rail transportation networks. To achieve the "80 in 40" goal, much of this infrastructure will have to be replaced or radically upgraded.

The transformation will touch everyone in the country. It will affect transportation, housing, work lives, and even personal habits. Some of these changes will be liberating; many will improve the quality of life. But change is never easy, and we should be wary of utopian visions that promise change without pain.

Of course, there is another way to cut carbon emissions—by shrinking the economy. Indeed, U.S. energy-related carbon dioxide emissions declined by 10 percent during the recession years of 2008 and 2009, and by more than 11 percent during the deep recession of the early 1980s.[12] But as a deliberate strategy, hobbling the economy is neither morally acceptable nor politically viable, and emissions reduction policies that do so have no chance of being sustained. As long as the economy continues to grow, the most important way to reduce emissions is through innovations in the way we supply and use energy.

Quantifying the innovation requirement in aggregate terms is actually a straightforward exercise. In this exercise, we consider not the G8 goal of 80 percent (which lacked a baseline year) but a similar, more precisely defined goal also articulated by President Obama: an 83 percent reduction in U.S. carbon dioxide emissions by 2050 relative to a 2005 emissions baseline.[13] Our key metric is the "carbon intensity" of the economy—that is, the amount of carbon dioxide emitted per unit of economic output. If we assume that the U.S. economy is going to keep growing, then carbon intensity must decline by more than 83 percent since the economy will be much bigger in 2050 than it was in 2005. How steep the decline will need to be depends on the rate of economic growth: the stronger the growth rate, the faster we will need to reduce carbon intensity.

For example, if the U.S. economy grows at an average rate of 2 percent per capita per year between now and 2050 (which would be neither particularly fast nor particularly slow by historical standards[14]) the carbon intensity of the economy would have to decline by about 7 percent per year to achieve the President's 83 percent reduction target.[15] We can disaggregate the 7 percent into two components, as shown in the following equation:

Rate of change in the *carbon intensity* of the economy = Rate of change in the *energy intensity* of the economy + Rate of *decarbonization* of the energy supply infrastructure

In other words, reducing carbon intensity can be achieved through a combination of (1) using fewer energy inputs to produce each unit of economic output (reducing energy intensity) and (2) reducing the share of carbon-based fuels in the energy mix ("decarbonization").

Aggregate reductions in energy intensity, in turn, can be realized through a combination of improvements in energy efficiency *within* individual sectors of the economy and shifts in the composition of the economy toward less energy-intensive sectors (for example, less steel production and more software design services). But here an important caveat is in order. If compositional shifts in the U.S. economy merely prompt the relocation of energy-intensive industries to other countries with high-carbon energy systems and less stringent emissions controls, there may be no climate benefit. For example, if a U.S. steel plant were to be closed while a plant elsewhere in the world is opened, the net impact on climate will be only the difference in emissions between the U.S. plant and the overseas plant, if any. A ton of carbon dioxide has the same impact no matter where it originates.

With this important proviso, the equation written above shows that decarbonization and reductions in energy intensity are substitutable: the more we can do of one, the less we will need of the other. But to achieve the 7 percent carbon intensity reduction target, *both* will have to improve much faster than they ever have before.

In fact, over the past 25 years, the carbon intensity of the U.S. economy has been declining at an average rate of about 2 percent per year. The trend is in the right direction, but the magnitude is far too small. Moreover, it turns out that almost all of this decline (more than 1.8 percent per year) was due to a decline in the energy intensity of the economy, achieved both within individual industries and as a result of structural shifts toward less energy-intensive industries. Over the same period the share of carbon-based fuels in the energy supply mix hardly changed at all—in other words, the rate of decarbonization was only slightly above zero. (See figure 1.2.)

Now suppose that in future years the country as a whole manages to accelerate the rate of decline in energy intensity from about 1.8 percent per year to about 3 percent per year. Such an improvement might seem modest, but in fact only a handful of American states have managed it in recent years. California, whose relatively aggressive policies have made it the poster child for energy conservation efforts, achieved only a 2.74 percent per year energy intensity decline over the past decade. (And, again, if energy intensity reductions were achieved merely by shifting carbon-intensive industries to other countries with high-carbon energy systems and then importing their products, nothing would be gained—we need real decreases.) In short, sustaining 3 percent per year over many years is a tall order.

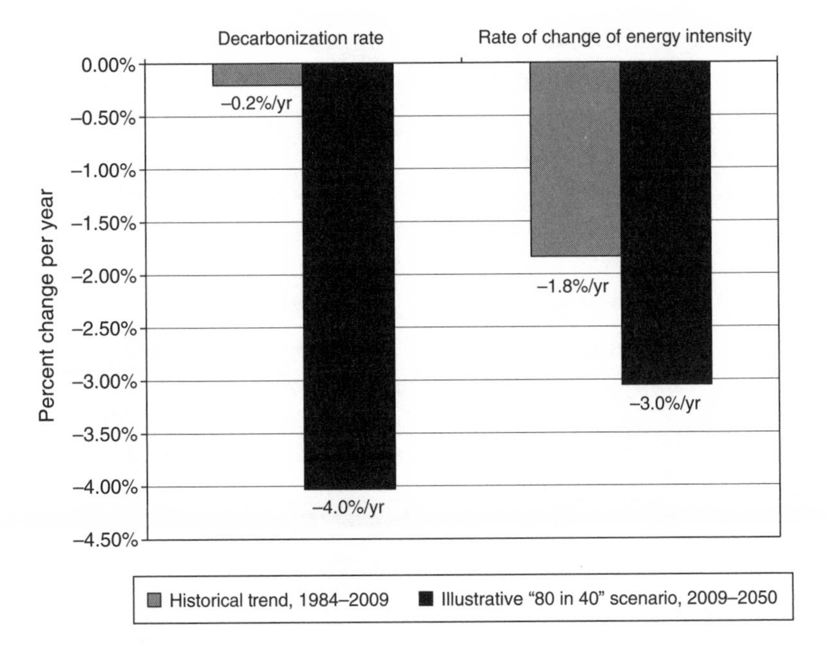


**Figure 1.2**

Rates of decarbonization and energy intensity reduction in the United States: (a) from 1984 to 2009; (b) for an illustrative "80 in 40" scenario, 2009–2050. *Source*: Lester and Finan, 2009.

Still, with a strong push from government and strong public support, there is reason to be optimistic that energy intensity reductions of this magnitude are achievable. In that case, the corresponding requirement for decarbonizing the U.S. energy system would be 4 percent per year (7%–3%) (see figure 1.2). What would it take to achieve that? The short answer is that it would take rapid rates of adoption of most if not all of the main low-carbon supply options: solar, wind, nuclear, geothermal, and advanced (low-carbon) biofuels, as well as carbon capture and storage (CCS). In one scenario for achieving a 4 percent per year decarbonization target, 120,000 megawatts of new low-carbon electric generating capacity would have to be added annually to the country's current installed base. That is a far faster pace of new installations than the United States has managed in the past. Over the last decade, for instance, the peak year for capacity additions of all kinds was 2001, during which 67,000 megawatts were installed (most of it natural gas). In 2009, a record-setting year for renewables, approximately 10,000 megawatts of new wind capacity and 450 megawatts of new solar photovoltaic capacity were installed around the country. Looking back a bit further, about 9,000 megawatts of new nuclear capacity entered service in 1986, the peak year for completing nuclear power plants.

In our scenario, wind and solar power would provide 40 percent of U.S. electricity in 2050, but would account for more than 60 percent of the total generating capacity at that time. (This difference is caused by the intermittent nature of wind and solar power. Even when they are sited in favorable locations, wind turbines and solar systems run a smaller percentage of the time than conventional generating technologies.) The remainder of the electricity required in our scenario would be provided by a combination of nuclear power and coal with CCS, together with smaller amounts of geothermal and hydropower. One way to reduce the required rate of new capacity additions would be to rely less on solar and wind and more on higher-capacity-factor nuclear and coal (or natural gas) systems with CCS. Another would be to implement demand-response programs designed to "shave" peak power requirements by promoting conservation at peak times and by shifting electricity demand to nonpeak periods. The deployment of grid-scale energy storage would also help. Taken together, these measures might reduce the annual requirement for new capacity additions by as much as 50 percent, but even this more modest goal would be difficult to achieve. Of course, the decarbonization target could itself be reduced below 4 percent per year. But that would mean having to raise the target for energy intensity reduction by an equal amount above 3 percent per year.

So we have a double tall order: any combination of reduced energy intensity and decarbonization that hits our carbon intensity target will be difficult to achieve in practice. We might be encouraged by the example of China, which has been building new power plants (unfortunately mostly coal plants) at a rate of about 100,000 megawatts per year for each of the last several years. But even aside from the (somewhat questionable) example of China, there are many reasons to believe that the target is not out of reach, if we put our minds to the task. There are enormous opportunities for innovation in energy efficiency and decarbonization that have the potential to yield a low-carbon energy system that is more reliable and quite possibly more affordable than the one we have now. We will shortly return to this theme and sustain it through the rest of the book.

But it is worth considering the alternatives. If the United States and other nations fail to stabilize the atmospheric carbon dioxide concentration at roughly the 550 ppm target level, and if the earth's climate system is no more forgiving than the best models predict, then humanity would have a different choice: we could either anticipate failure and invest ahead of time in actions that might help us adapt to climate change with less pain, or we could simply decide to live with the consequences of failure when they arise. (In fact, even if the United States sets a course to achieve "80 in 40," investments that would enable us to adapt to climate change—to "manage the unavoidable" in presidential science advisor John Holdren's words—would still be prudent, since some warming would continue to occur—albeit on a much less disruptive and dangerous trajectory.)

## The Essential Role of Electrification and "Bridges" to Get There

The scenario described in the previous section points to another important feature of the transition: decarbonization will mean a much larger role for electricity in the U.S. energy system. Many of the most promising low-carbon technologies are much better suited to electricity generation than to the direct production of heat or other forms of energy. Nuclear, wind, and solar technologies as well as CCS fall into this category. Natural gas (without CCS) and possibly also "second generation" biofuels (i.e., biofuels made from wastes or other, less energy-intensive crops than corn) may help build a bridge to a decarbonized, electricity-centric energy system in the next three decades or so, but these bridging fuels will eventually become unsustainable sources of carbon emissions themselves.

Electricity consumption as a share of overall energy use has been growing steadily ever since Thomas Edison launched the era of commercial electric power at the end of the nineteenth century. Today 41 percent of America's primary energy supply (domestic and imported) is used to produce electricity, compared with just 18 percent fifty years ago. For households, businesses, and industries, electricity has become the preferred energy carrier for a vast range of uses. The information technology revolution and the devices it has spawned have further accelerated this trend. Digital data centers, for instance, which barely existed two decades ago, accounted for 1.5 percent of U.S. electricity consumption in 2007—

more than all the electricity used by the nation's televisions—and their consumption was expected to double by 2011.[16]

Aggressive efforts to decarbonize the U.S. economy will likely accelerate the trend toward electrification. In our scenario, electricity use would nearly triple between now and 2050, even with substantial efficiency improvements. Transportation accounts for nearly all of this increased demand, as the vehicle fleet shifts away from petroleum. Our scenario stands in stark contrast to business-as-usual scenarios that predict only minimal changes in the transportation sector and slow growth in electricity demand in the coming decades.[17]

Biofuels do not change that conclusion. Ethanol derived from corn is the dominant biofuel in use today, and it will likely be the primary fuel used to meet a federal renewable fuels requirement in the near term. Unfortunately, the carbon footprint of corn ethanol is greater than that of gasoline, once the whole fuel cycle is taken into account, including the energy inputs to grow the corn and convert it into fuel as well as the indirect impacts on land use and food markets.[18]

"Second generation" biofuels would use plant waste or other uncultivated or lightly cultivated source material, thereby avoiding the most energy-intensive steps used in making corn ethanol. If these technologies mature quickly, they have the potential to cut transportation sector carbon dioxide emissions significantly, without requiring a complete overhaul of the petroleum-based infrastructure. Progress toward commercializing second-generation biofuels has been slow, however. And on a 40-year time horizon, using the modest economic growth assumptions in our scenario, emissions from these kinds of biofuels would still breach our target. Third- and fourth-generation biofuels that would be truly carbon-neutral (in the sense that no more carbon dioxide would be released during combustion than would be absorbed from the atmosphere to produce the fuels in the first place) are still in the early research phase. Their main contribution will come, if at all, only after 2050.

Natural gas already plays a major role in America's energy supply mix, and its contribution will certainly grow in the medium-term before it becomes emissions-limited, like biofuels. The supply outlook for natural gas has shifted dramatically in recent years as new horizontal drilling techniques and hydraulic fracturing technologies pioneered by independent drillers in the 1980s and 1990s to extract gas from shale formations have been adopted by the rest of the industry. The result has been a large expansion in shale gas production, from negligible levels a decade ago to 25 percent of total domestic gas production today.[19] Adding to the optimism, natural gas prices have dropped sharply, from $8 per thousand cubic feet at the well-head in 2008 to about $4 today.

The U.S. Energy Information Administration (EIA) now estimates that shale gas accounts for one-third of all technically recoverable U.S. natural gas resources and could account for nearly half of all domestic gas production by 2035.[20] At this level, shale gas would more than offset the expected decline in output from conventional gas resources. Much of this new production will come from the Marcellus shale, a formation that underlies New York, Pennsylvania, and other eastern states that have not experienced much oil and gas development in the recent past. The EIA has begun to refer to shale gas as a "game changer."[21]

Increased use of natural gas would certainly reduce carbon dioxide emissions in some applications. The most significant benefits could be achieved by replacing old, relatively inefficient coal-fired power plants. (This would have the added benefit of reducing other emissions from the coal plants, including sulfur and nitrogen oxides, particulates, and mercury.) The quickest and least expensive approach to doing so is to use existing but currently underutilized gas plants for this purpose. Further displacement of coal with new combined-cycle gas-fired power plants might also be economically attractive in some cases. This upgrade could reduce carbon dioxide emissions per kilowatt-hour by 50 percent or more. Replacing the entire existing fleet of coal power plants with modern gas-fired plants would reduce U.S. carbon dioxide emissions by 22 percent overall.

It is far from clear, however, that the new shale gas discoveries will be large enough to support an expansion of this magnitude. To replace all of the coal currently used to generate electricity, for example, domestic natural gas production would have to increase by 54 percent. Another possible market for low-cost natural gas would be as an alternative motor vehicle fuel. Today, the transportation sector accounts for less than 0.2 percent of natural gas consumption (mostly for buses that run on compressed natural gas, or CNG). Replacing one quarter of total current gasoline consumption with CNG would increase domestic gas consumption by 21 percent. An alternative approach would be to convert the gas to liquid fuels such as methanol or diesel, although this would probably be less efficient (and might even lead to increases in greenhouse gas emissions). Not even the most optimistic estimates suggest that new shale gas production could support this level of increased demand while also compensating for the continued decline in conventional gas production.

Shale gas will probably help to moderate natural gas prices. It has the potential to be a useful bridge to the low-carbon economy. But it is not likely to change the energy innovation challenge fundamentally. Ultimately, America will need even lower-carbon (and more abundant) fuels than natural gas, if energy is to be available on the scale that Americans have come to expect.

## Mitigating Climate Change and Other Energy Policy Goals

Climate change is the most serious and most complex energy challenge that America faces when all the scientific, technological, economic, and political aspects are considered together, as they must be. But as we noted at the outset, climate change is not the only serious energy challenge the United States confronts this century. Reducing the nation's dependence on oil imports; mitigating other environmental impacts of energy supply and use; developing energy industries as a way to create domestic jobs; ensuring affordable energy prices so as to stimulate economic growth and competitiveness; strengthening the reliability of energy systems and reducing their vulnerability to accidents and sabotage—all of these goals demand attention and at various times each has dominated the national energy policy debate. We seek, to the maximum extent possible, an innovation system that can make progress on these other goals even as it helps mitigate climate change. At the same time we recognize that there will sometimes be conflicts that force us to choose among goals.

The most prominent stated goal of U.S. energy policy since the early 1970s has been to cut oil imports. Despite many exhortations and interventions by the president and Congress, the United States has made no significant progress in this regard—indeed, crude oil imports have continued to rise. They are now two-and-a-half times greater than they were in 1973. Moving to a more efficient, less carbon-intensive energy system could help to reverse this trend. Electrifying the vehicle fleet, for example, would reduce our dependence on hydrocarbon resources from the Middle East and other volatile parts of the world. As the electric utility expert

Paul Joskow has noted, the electricity sector is a model of energy independence. (Although some of the uranium used in U.S. nuclear power plants is imported, uranium supplies are less concentrated in volatile regions of the world, and uranium can also be stockpiled more easily than oil.[22]) However, if the electricity system relied on carbon-intensive resources, such as coal or oil extracted from tar sands without carbon capture, electrifying transportation could accelerate climate change.

Mitigating energy-related environmental impacts and public health risks other than climate change has been another important thread of recent energy policy. Battles over mountaintop coal removal, coal mine safety, and deep-sea oil drilling are ongoing, as are efforts to address the adverse local and regional health effects of mercury, particulates, and other pollutants from coal-fired power plants. To the extent that mitigation of climate change induces a shift away from fossil fuels it will also produce co-benefits with respect to these local and regional impacts. But new forms of energy production will raise a new set of local and regional concerns. The siting of large solar farms in pristine Southwestern deserts and the use of hydraulic fracturing techniques to extract natural gas on a large scale have already sparked conflict.

Any large-scale shift in energy supply and use would also have economic implications that need to be considered. Scaling back domestic coal use will reduce power sector carbon emissions, but the economic impact on mining companies and the railroad industry (where coal accounts for nearly 50 percent of all freight tonnage) could be severe. Even if such a shift were stretched out over many years, employees and investors in these industries, as well as the communities that depend on them, are likely to oppose it. Financial compensation for such adversely affected interests will in all likelihood be a significant cost of the energy transition.

At the same time, the energy transition will undoubtedly yield new jobs. An obvious area for job growth is retrofitting existing buildings for energy efficiency. Such jobs would be widely distributed—they would exist wherever there are older buildings—and many of them would involve construction trades that have contracted drastically since 2008. Installing new, low-carbon power plants and smart-grid improvements would likewise create new employment opportunities in construction, engineering, and related fields. The location of manufacturing jobs for components for these systems is less clear; some manufacturing will likely be located near the target markets but some will probably occur in concentrated "clusters" that serve customers globally. Not all of these clusters will be in the United States, but the United States is well-situated to capture a sizeable share of the global value chain in some of the most important low-carbon industries.

The net effect of the energy transition on aggregate employment, then, is uncertain. One of the strengths of a dynamic economy like ours is that it can accommodate shifts between industries over time as waves of innovation break again and again over the economic landscape. We expect the twenty-first century energy transition will be comparable to earlier transitions (albeit faster), such as the shift from a wood-fueled economy to a coal-fueled economy in the nineteenth century, and the subsequent shift to a petroleum-based economy in the twentieth century. A low-carbon economy is unlikely to be either the "green jobs" panacea or the job killer it is sometimes made out to be, although some places and some occupations will undoubtedly benefit (and suffer) more than others.

Affordability is an even more important goal of energy policy, in our view, than employment. For many businesses, energy costs are an important factor in economic competitiveness. Energy costs are also highly visible to individual citizens and sometimes have become a major political issue. Many decarbonization options are very expensive today. Capturing and sequestering the carbon dioxide emitted by coal-burning power plants—a technological feat that has yet to be demonstrated at full scale—would almost double the cost of producing electricity from coal, for example.[23] Similarly, the cost of generating electricity using solar photovoltaic technology installed at residences (without subsidies) is several times the average price paid by electricity consumers today. Not all low-carbon technologies are expensive. When sited in the most favorable locations, where the wind blows steadily and strongly, modern wind turbines can produce power at a cost that approaches parity with fossil fuel generators. Many energy efficiency measures are cost-effective today without subsidies as well. Still, if the coal, oil, and gas that today provide 83 percent of U.S. primary energy were somehow replaced at a stroke with a combination of efficiency measures and low-carbon supply options at today's costs, the nation's energy bill would rise sharply.

Reliability is as important as affordability. Americans have come to expect that they can flip a switch or turn the ignition whenever they want and wherever they are. They are not accustomed to planning their days around power cuts or long lines at the gas pumps, as many people elsewhere in the world do. The low-carbon transition is likely to create new reliability challenges. Energy from the wind or the sun, for instance, is intermittent and must therefore be balanced by other sources on a minute-to-minute basis. The "smart grid" will depend on complex information technology systems that could have bugs or breakdowns. Energy prices that change at different times of day, which are necessary to make the smart grid work, might disrupt some existing routines. On the other hand, a greater diversity of supply resources and more sophisticated controls should, on the whole, make the low-carbon energy system of the future more reliable and robust than today's system.

In sum, there are good reasons to believe (and better reasons to hope) that low-carbon options exist that will affordably and reliably meet the energy needs of a growing twenty-first century economy. The United States can unlock their potential without, at a minimum, exacerbating any of the other problems that concern contemporary energy policy-makers. If luck is with us, we might be able to make simultaneous progress on many fronts.

## The Real Energy Innovation Opportunity: Doing Much Better Using What We Already Know

An industrial transformation on the vast scale of the forthcoming energy transition will not yield to a single solution. It will require instead a broad portfolio of options that are steadily refined and winnowed over time. The next few decades are critical. Forty years might seem like a long time, but the energy system is so big, so complicated, so important, and therefore so slow to change that we cannot afford to wait and see if potential breakthroughs that are still at the laboratory stage today (much less concepts that have not yet been tested in practice at all) will mitigate climate change. However promising they may sound in theory, it is implausible that such inventions will be deployed by midcentury on a scale large enough to have a major impact. Instead, most of the heavy lifting will have to be done by technologies whose basic scientific and engineering characteristics are already well known, but whose performance has the potential to be improved along several dimensions through further refinement both in the lab and in the field. This is the real energy innovation opportunity.

Our argument against the breakthrough thesis is based, first of all, on experience. It typically takes at least several years to move a new energy technology from "proof of concept" in the laboratory to full-scale demonstration, and many more years or even decades before that technology can achieve significant market penetration. The first modern solar photovoltaic (PV) device was developed at Bell Labs in 1954; yet more than half a century later, PV systems—most of them employing the same silicon-based technology demonstrated in that first device—still only account for one hundredth of 1 percent of total U.S. electricity generation. Wind power is an even older technology. Like PV it has received generous R&D subsidies and occasional market support since the 1970s. Today wind accounts for just 2 percent of total U.S. electricity generation. Even nuclear energy, the fastest-growing new power source of the last half century, took 35 years following its initial demonstration in 1942 to gain a 10 percent share of the electricity market, and more than another decade to reach the 20 percent mark—again in spite of a strong national push to deploy nuclear power during much of that period.

Why does energy innovation take so long? Some reasons are not specific to energy. Breakthroughs in any field must be refined before they are embodied in viable products and kinks in the production process must be worked out; all of this takes time. Then users must become familiar and comfortable with new products, a process that sometimes entails institutional changes, new business models, and evolution of social norms—this too takes time. In addition, much of the existing energy infrastructure (e.g., power plants, refineries, transmission networks, and residential and commercial buildings) has a lifetime of many decades. This mammoth infrastructure, like any long-lived capital, turns over only slowly.

Factors particular to the energy sector also tend to slow the pace of change. First, energy innovations frequently involve complex systems. This means that before a new technology can be adopted at scale several related innovations are also required. Large-scale deployment of wind and solar systems, for example, will entail solutions to a number of associated grid-related challenges, such as developing and implementing electricity storage technologies, more accurate forecasting tools, new load management procedures, and novel simulation and control technologies.

New nuclear reactor technologies may require a cluster of innovations in the nuclear fuel cycle to support them. The mass deployment of electric vehicles will require not only advances in battery technology, but also a new battery recharging infrastructure, repair and maintenance capabilities, and significant modifications to the vehicle manufacturing supply chain.

Because of the long lead-times involved, the country cannot afford to place its hopes on achieving a truly radical technological transformation of the energy infrastructure during the next few decades. Rather, energy innovation during this period should focus on bringing down the costs, improving the scalability, integrating diverse components, and reducing the environmental impacts of energy systems that not only are identifiable today but in many cases are already in the marketplace or under development. These improvements will require sustained effort and creativity in laboratories, at production sites, on user premises, in executive suites, and in the halls of government. Better components, improved materials, new organizing concepts, novel business models, and more are required. There is an enormous innovation agenda that does not depend on a "moonshot" mentality.

Even so, research aimed at radical breakthroughs still warrants support and attention. Even if emission reduction efforts over the next few decades are successful, the ultimate goal of stabilizing atmospheric greenhouse gas concentrations will take longer to achieve and will require further deep cuts. Breakthroughs in science and engineering made during the present half-century—whether in energy storage, biofuels, solar energy, or any other field—will most likely make major contributions in the second half of the twenty-first century. In the best case it will be possible to overcome longstanding physical limits on the performance of energy transport, storage, and conversion technologies. Advances of this type are unlikely in the absence of sustained research on fundamental problems in fields such as thermal and electrical energy transport, new classes of materials, and catalysis. Although the benefits of such advances may not be observable at scale for some decades, the research needed to realize them—across diverse fields including life sciences, materials science, and computational science, as well as the energy sciences—must be pursued vigorously today.

## Conclusion: From Innovation Agenda to Innovation System

The facts, deductions, and grounded assertions in this chapter establish an innovation agenda and provide an outline of the requirements that it must fulfill for the United States to get to "80 in 40." Any combination of reduced energy intensity and decarbonization that does not rely on wishful thinking about heretofore-unknown "game-changers" and that adds up to a 7 percent annual reduction in carbon intensity, sustained through 2050, will do.

Making such statements is easy. It is much harder to conceive an energy innovation system that can carry through such an agenda and to envision how that system can emerge from the one that the United States has now. Conveying our insights about those problems is the real burden of this book. The next chapter articulates the analytic foundations and governing principles that will guide us as we grapple with them.

# 2

# An Energy Innovation System That Works

Chapter 1 made the case that a fundamental transformation of the U.S. and global energy systems needs to be well underway within the next two to three decades. Success in bringing about this transformation will require greatly accelerating the pace of energy innovation. Innovation to improve the portfolio of low-carbon options offers the only palatable path forward to meet the energy needs of an expanding U.S. economy while substantially reducing the amount of greenhouse gases released to the atmosphere.

We envision three waves of innovation. The first wave should focus on making energy use more efficient, beginning on a large scale right away. The second wave should aim to decarbonize energy supply, scaling up between 2020 and 2050. The third wave should be driven by yet-to-be realized breakthroughs that would play a major role in the energy system in the second half of this century. All three waves will involve a wide array of innovations, not only in technology, but equally in business, societal institutions, and public policy.

We need the rest of the world to take these steps, too, but aggressive action on our part is necessary for global success. This is not only because of the magnitude of U.S. greenhouse gas emissions, both now and in the past, but also because the U.S. economy is the innovation engine of the world. Throughout this nation's history, and especially in the past century, Americans have invented new products and services, new systems of production, new ways of organizing business and finance, and new modes of collective action and public deliberation. Not every one of these innovations has worked out as well as its proponents might have hoped, but in the aggregate the benefits have been enormous—and have greatly outweighed the costs. Of course, the United States does not have a monopoly on innovation—great ideas have always come from many places, and this is likely to be more and more the case in the future. (In fact, one of the things that the United States needs to get better at is absorbing innovations made elsewhere.) But, for all the talk of American decline, the United States remains the world's preeminent source of innovation. Mitigating climate change will require so much innovation on a global scale that there might not be enough of it to get the world where it needs to go if the United States is not a big part of the effort.

Why has the United States been so good at innovation? It isn't that Americans are smarter or more creative or harder-working than people anywhere else. It isn't that business strategists in this country are more gifted or that investors are savvier than anywhere else. It isn't that U.S. legislators craft more cunning policies and government agencies carry them out more effectively than anywhere else. It does have something to do with the fact that the U.S. economy is huge and complex. But perhaps more than that, it is because the United States has a unique set of market and non-market institutions that fit together into a *system* that has been extremely innovative on many fronts over a long period of time—from agriculture to automobiles to information technology to the revolutionary nano-, neuro-, and bio-technologies that are just emerging on the scene today.

This innovation system is complicated. It includes public and private research laboratories; small entrepreneurial firms; large, mature firms; financial intermediaries ranging from huge banks to individual "angel" investors; educational institutions, from schools to community colleges to universities; local, state, and federal government agencies; and innovation users of many different types. These organizations and individuals are knit together by a set of beliefs, norms, incentives, and laws that give each a productive role to play. A lot of competition exists, but so does mutually beneficial cooperation. The system doesn't work perfectly. But on the whole the American innovation system has done a remarkably good job of uncovering new opportunities and developing and commercializing them.

The time has come to fully extend America's innovation system to meet the energy challenges of the twenty-first century. For a variety of reasons the United States has been less innovative in the energy sector than in others. Some of the reasons have to do with the energy market and the consumers and suppliers who inhabit it. Others have to do with the government, especially the federal government, which has performed poorly in energy innovation, despite occasional bursts of action in the wake of recurrent crises. But mostly it is because we really haven't tried. Energy has been available, reliable, and cheap—why change?

The time has come to try, for all the reasons discussed in the previous chapter. We know that market forces alone will not drive innovation in response to the threat of climate change, or at least not fast enough to avert potentially unmanageable outcomes. Environmental costs, especially those far away in space and time, are not now priced into energy sales. And even if they were, additional steps would still need to be taken—for reasons we discuss in this chapter—to break out of the stagnation that has characterized the American energy innovation system for so long. Only the federal government can take those steps, although state and local governments have important roles to play as well.

Our emphasis on governmental actions is not intended to diminish the role of the private sector. The energy transition will primarily result from countless private decisions on energy supply and use, shaped by the entrepreneurial actions of private innovators. Private innovators are at the heart of the U.S. innovation system. But private innovators respond to signals sent by the markets in which they operate. And markets, especially energy markets, reflect the broader social context in which they are embedded. That context has so far failed to send the signals that would accelerate low-carbon energy innovation.

This chapter begins with a discussion of the innovation process and the innovation system as concepts, both in general and for energy in particular. We turn next to the U.S. energy innovation system as it works in practice, concentrating on the U.S. Department of Energy at the federal level. As we do so, we identify a set of weaknesses in the system's operation. We then take up a series of proposals that have been offered to address these weaknesses. In our view, these proposals do not fully meet the challenge. The conclusion of the chapter puts forward a set of principles for building an energy innovation system that would do so.

## The Innovation Process and the Innovation System

It is impossible in this short space to do justice to the large and growing literature on innovation systems. A key insight of this literature is that innovation is not solely, or even primarily, the work of the lone creative genius who experiences a flash of insight. The seeds of innovation more often germinate from collaborations among large interdisciplinary teams or between product designers and prospective users than in backyards and garages. A wide range of institutions may be involved. Moreover, the initial insight is just the first step (albeit an important one) of many in the innovation process. Whatever its source, if a new idea is to create value it must be reduced to practice: that is, converted into a product or a process or a service that works. This stage, too, may involve many people and a variety of institutions, including pioneering small companies, large multisectoral collaborations, and groups of users.

After it has been reduced to practice, the innovation must then be road-tested by its users, to show that it is economically viable and that there is demand for it. Then, to have real impact, it must be "scaled"—that is, adopted by a significant fraction of the population of potential users. Most innovations undergo continued refinement even after they have been deployed at scale. Many institutions of different kinds are involved in these later stages of the process as well. Also, operating in the background of all these activities are laws and policies—like those governing intellectual property, taxation, and trade—along with regional and national cultures that may influence the pace and direction of innovation, especially in innovation hotspots like Silicon Valley.

The process of innovation is complex and rarely unfolds in a linear manner, but we find it is helpful to distinguish four basic stages in the progression from idea or concept to large-scale deployment (see figure 2.1):

- Discovering and developing new possibilities ("option creation")
- Ascertaining the viability of these possibilities in practice ("demonstration")
- Reducing costs and risks and making other improvements during initial take-up of the innovation in the marketplace or by noncommercial users ("early adoption")
- Continuing refinement of the innovation in large-scale use ("improvement-in-use")

Each of these stages is worth considering in somewhat greater detail.

*Option Creation* The goal at this stage is to open up a broad range of innovation pathways by encouraging experimentation with new ideas

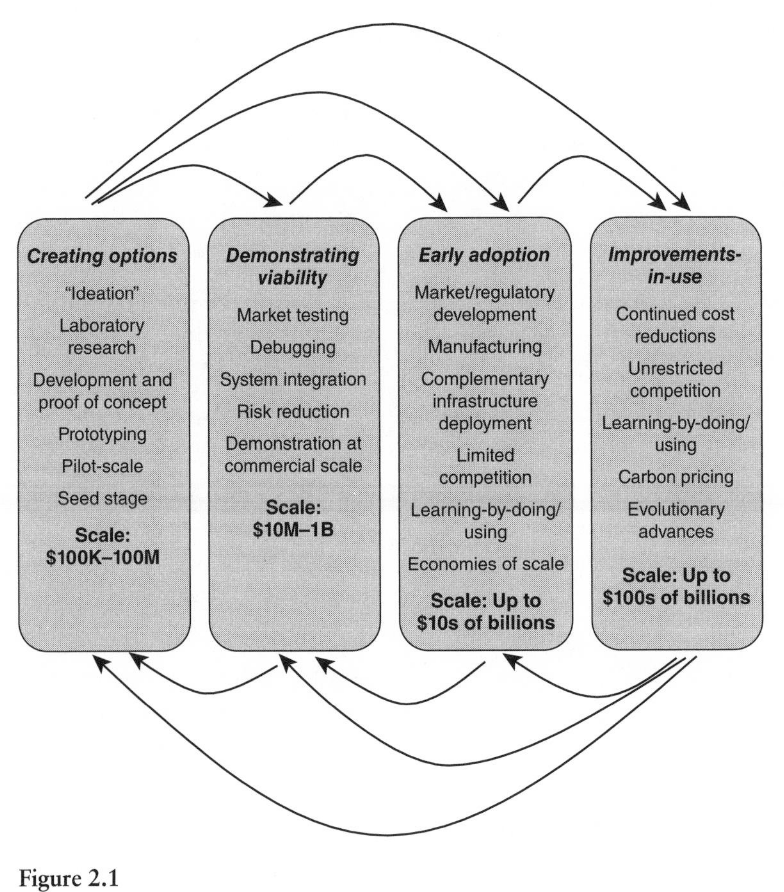


**Figure 2.1**

The four stages of energy innovation.

and concepts, by attracting new entrants to participate in the process, by ensuring that knowledge about the options being explored is generated transparently, and by guaranteeing broad access to that knowledge. An important activity at this stage is "proof-of-concept" testing to establish that there are no technical showstoppers that would prevent the new concept from being realized in practice. Option creation is closely associated with R&D, but the two are not synonymous. Although big new technical ideas often grow out of organized, well-executed R&D programs, ideas for new products and services—including new business models—can arise anywhere. (And, reciprocally, much R&D is not devoted to option creation but to supporting other stages of the innovation process.) From the perspective of the innovation system as a whole, a key goal at the option creation stage is to maintain and expand the technical communities and 'public spaces' that facilitate the generation of new knowledge and concepts.[1] Public funding must play a large role in this stage.

*Demonstration* In the demonstration stage the primary objectives are (1) to enable technology providers, investors, and users to obtain credible information about cost, reliability, and safety under conditions that approximate actual conditions of use, and (2) to make this information broadly available (consistent with proprietary constraints). Put differently, the goal of demonstration is to reduce technological, regulatory, and business risks to levels that would allow private investment in the first few commercial projects. This entails building, operating, and debugging full-scale prototypes. Private innovators and their investors assume an increasing share of costs and risks in the demonstration stage, compared to the option creation stage—in some cases they assume all the cost and risk. But the time horizon may be too long and the risk level too high for private investors to be willing to underwrite demonstrations for complex, large-scale technologies. Other important tasks at this stage include settling on standards and infrastructure requirements and identifying key legal and regulatory barriers that will need to be overcome for widespread use.

*Early Adoption* Early adoption involves the most forward-looking users, or perhaps those with the strongest need to use the innovation. The main goals in this stage are market development, accelerated learning, and early deployment of the infrastructures needed for scale-up. Innovators establish manufacturing and distribution capabilities and other key parts of the supply chain, while the early adopters (sometimes known as "lead users") also play a key role, providing feedback that allows valuable features to be enhanced and practical problems to be sorted out. Proprietary knowledge about processes builds up during early adoption and unit costs generally come down (although aggregate costs increase because the number of units is growing rapidly).

*Improvement-in-use* The market and regulatory environments in this final stage settle into stable and predictable patterns. But it would be a mistake to think that innovation stops once these patterns are established. Designs continue to be refined, production systems and business models are improved, and the behavior of customers comes to be better understood. Frequently the cumulative impact of evolutionary improvements to a technology during its lifecycle in the marketplace greatly exceeds the performance gains achieved when the technology is first brought to market.

Although we have laid out the four stages of innovation—option creation, demonstration, early adoption, and improvement-in-use—in a sequence, this is not a simple linear system. Knowledge flows between these activities in both directions (as represented by the arrows in figure 2.1), and in reality the boundaries between them are blurred.

### *Energy* Innovation: System and Process

America's national institutions and dynamic regional clusters have proven to be great assets for innovation across all of these stages in many industries. That has been especially true in new industries, where innovation involves the creation of fundamentally new products and services. These assets can be applied to energy innovation, too. But energy is different, and the United States will need to take some important steps in order to leverage the strengths of its innovation system in this domain.

The first thing that any potential innovator needs to know about energy is that it is a commodity. Users care about the services energy provides, not about where the energy comes from. Mainly they care about reliability and cost. Highly optimized systems, owned and operated by well-financed, well-established, and politically influential incumbent firms, have evolved over more than a century to supply this commodity. This is the toughest kind of innovation environment. The challenger faces stringent, non-negotiable requirements on cost, quality, and reliability from the outset—along with competition from entrenched interests with deep connections on Main Street, on Wall Street, and in Washington.

To put it another way, the energy sector is deeply embedded in the fourth stage of the innovation process. This is not to say that energy innovation has stopped: on the contrary, market drivers—including particularly the demand for new energy supplies—continue to produce advances, such as those that have recently made it possible to access deep offshore oil resources and extract natural gas from shale formations. By contrast, innovation aimed at reducing greenhouse gas emissions or at furthering other goals of energy policy are not being driven (or are being driven only weakly) by the market. The impetus for these kinds of improvements will have to come from outside the existing, well-established system. These observations have implications not just for the first stage but for all four stages of the innovation process. An innovation system that develops new low-carbon energy technologies but that does not result in these technologies being adopted on a large scale will not have succeeded at all. The whole point is to achieve scale.

*Option Creation* Large incumbent energy firms have little incentive to create new options, especially in the absence of any signal from the government that climate change or other energy policy goals will be taken seriously. That means that option creation must move forward primarily in organizations that are sheltered from market pressures, such as university and government research laboratories or in collaborations that cut across institutional and sectoral boundaries. Venture capital-funded start-ups may also play a role in creating options in some segments of the energy field, but the enormous scale and long time horizon of many energy systems limits the range of innovations that such start-ups can undertake.

*Demonstration* For smaller-scale energy innovations, including many energy efficiency innovations as well as some technologies related to distributed electricity generation and the "smart grid," private investors may be willing to assume all of the costs and risks of the demonstration stage—particularly if the regulatory environment is hospitable. But demonstrating innovative central-station power plants, systems for carbon capture and sequestration, and other large-scale innovations may cost hundreds of millions of dollars or more. Licensing and other regulatory uncertainties for these innovations are also large. In such cases, private firms are unlikely to move forward unless public institutions share the costs and risks of demonstrating the new technology.

*Early Adoption* This stage of the innovation process may be the most challenging of all for many energy technologies. Customers clearly care about energy reliability and affordability, but we also know that there is a lot of learning that makes innovations more reliable and affordable as they go through the early adoption stage. Many low-carbon innovations will still cost more than incumbent fossil fuel technologies for some time even after they have been demonstrated. Innovators therefore may need protection from incumbents during this stage and perhaps also public support for a limited time and to a limited extent in order to build their proprietary capabilities, even as they compete intensively with one another. But to create the right incentives for further cost reductions, these protections and subsidies must be phased out by the end of this stage.

*Improvement-in-use* Once energy efficiency and low-carbon energy supply systems have been road-tested and debugged, they should compete with one another and with incumbent technologies on level terms. Predictability, rather than stability, is the key attribute of the institutional environment in this final stage of the innovation process. Predictability is important because the climate change challenge demands a continual ratcheting down of emissions. Energy innovation policy must avoid recreating an entrenched system with a new set of incumbents. The goal instead must be to build a system in which continual innovation becomes the new normal.

## The Department of (Not So Much) Energy

Differences between energy and other sectors, along with the urgency and enormity of the climate challenge, argue for an active government role in energy innovation across all stages of the innovation process. To be more precise, government will have to play a carefully calibrated *set* of roles that are different at each stage, so that big firms, small firms, investors, universities, the legal system, and other key institutions can fit together into a more effective energy innovation system. The U.S. government has had a Department of Energy (DOE) for more than thirty years. But, ironically, energy innovation has never been DOE's primary focus, and many of its attempts to foster it have gone awry.

DOE was created in 1977 in response to the Arab oil embargo and the ensuing energy crisis. It brought together programs that had previously been distributed across several agencies, including the Atomic Energy Commission (AEC), the National Science Foundation, the Department of the Interior, and the Environmental Protection Agency.[2] In spite of its name, nuclear weapons and defense programs inherited from the AEC (which was itself the successor of World War II's Manhattan Project) have always taken a large share of DOE's attention. Other important non-energy programs lodged within DOE include fundamental research in high energy and nuclear physics, much of which also grew out of the Manhattan Project. Cleaning up the messes left behind by the Manhattan Project and Cold War nuclear complex is a third major area of activity. In fiscal year 2008 (the last year before the department's budget was temporarily transformed by stimulus funding), defense and environmental remediation programs accounted for 62 percent of a total DOE budget of about $24 billion. Energy research, development, and demonstration (RD&D) activities (including the Basic Energy Sciences program office) accounted for $5.7 billion or about 24 percent of the overall DOE budget.[3]

The history of DOE's energy RD&D activities is a history of boom and bust, reflecting the on-again, off-again nature of the federal government's engagement with energy issues. In real terms, the energy RD&D budget peaked in DOE's first year of existence. It declined rapidly during the 1980s and then declined further during much of the 1990s. By 1998, DOE spending on RD&D had fallen to about one-fifth of its level in 1978. Beginning in 2000, the RD&D budget began to increase slowly, so that today it is back to where it was in the early 1980s, or about half of the 1978 peak in real terms.[4] Cumulatively, the department has allocated about 29 percent of its total energy RD&D spending over the last thirty-plus years to fossil fuel technologies, 19 percent to energy efficiency, 18 percent to renewables, 15 percent to fusion, and 11 percent to fission. There have been significant shifts in the amount of funding directed to each of these fields during DOE's existence, as well as shifts in emphasis between applied technology development and long-term, high-risk research.[5]

Historically, DOE's network of national laboratories (the largest of which are Los Alamos, Sandia, Livermore, and Oak Ridge) has received about two-thirds of the department's RD&D investment for basic and applied science and engineering in both energy and non-energy fields. In recent years, DOE has launched several new initiatives designed to strengthen its energy RD&D portfolio, including three Bioenergy Research Centers and 46 Energy Frontier Research Centers. These centers are based at both universities and national labs and have typically involved collaborations between them. Each center is conducting fundamental research directed toward solving "grand challenge" problems in fields identified by the scientific community.

Congress has also approved three larger-scale "Energy Innovation Hubs," which bring together researchers from universities, private industry, and government laboratories to address critical national needs. The three hubs funded thus far are working on modeling and simulation for nuclear reactors, energy efficient building systems design, and fuels from sunlight. In each case the goal is to bridge the gap between fundamental scientific breakthroughs and industrial commercialization. Finally, a new agency, Advanced Research Projects Agency–Energy or ARPA–E (modeled on the Defense Department's well-known Defense Advanced Research Projects Agency or DARPA), has been established within DOE to fund small teams to develop "transformational" energy technologies that have the potential to have a large impact on national needs but that are too risky to be funded by private industry alone.

In addition to providing direct support for RD&D, DOE administers other programs and policies that stimulate energy innovation indirectly. Examples include the ENERGY STAR voluntary labeling program for appliances and buildings, the Building Technology Assistance Program, the Clean Cities Program, and the Federal Energy Management Program. DOE oversees a number of regulatory functions as well, including setting and revising energy efficiency standards for lighting and household appliances. Yet another important initiative is the Title XVII Loan Guarantee Program, established in 2005, which authorizes the Secretary of Energy to provide loan guarantees for up to 80 percent of the cost of projects that employ innovative technologies to avoid, reduce, or sequester greenhouse gases or other air pollutants.

It is not a simple matter to assess DOE's record. A series of troubled demonstration projects and programs casts a long shadow over it. These include the Clinch River Breeder Reactor Project, the synthetic fuels program of the late 1970s and early 1980s, the Yucca Mountain nuclear waste repository project, and the still unfolding FutureGen demonstration project for carbon capture and sequestration. Many of the problems within these projects and programs were not of DOE's making but rather were imposed on it by interest-group politics and interregional conflicts that were played out through Congress and successive administrations. But other chronic problems are internal to the department: a systematic tendency to underestimate project costs, inefficient business practices, and an impulse to push projects, technologies, and subsidies long after their unsuitability has become obvious. One study of six large federal technology commercialization projects, mainly in the energy field, concluded: "The overriding lesson from the case studies is that the goal of economic efficiency—to cure market failures in privately sponsored commercial innovation—is so severely constrained by political forces that an effective, coherent national commercial R&D program has never been put in place."[6]

To be sure, not all of the news has been bad. A decade ago a committee of the National Research Council assessed DOE's portfolio of energy efficiency and fossil energy research programs between 1978 and 2000 and concluded that these programs had "produced economic benefits, options for the future, and knowledge benefits." Although the committee was not always able to separate the DOE contribution from that of others, it judged that "the net realized economic benefits in the energy efficiency and fossil energy programs were . . . in excess of the DOE investment." Some of the energy efficiency investments were found to have paid off particularly handsomely.[7]

On balance, we take a rather dim view of DOE's past contribution and future potential to contribute to accelerating energy innovation, especially in the critical demonstration and early adoption stages of the process. In later chapters we propose to build on DOE's strengths, notably in option creation and in R&D that supports other stages of the innovation process, while devising new institutions that compensate for DOE's weaknesses.

## Beyond DOE: Federal and State Government Agencies in the Energy Innovation System

Energy is so ubiquitous and so important to today's society that many government agencies other than DOE, at the state as well as the federal level, have an influence on energy innovation. In fact, from a budget standpoint, energy-related programs housed in other federal agencies—like the Department of Agriculture and the Department of Transportation—as well as tax incentives administered by the Treasury, dwarf DOE's energy activities. According to one recent estimate, only 17 percent of total federal energy-related spending is allocated to R&D, while more than 80 percent is in the form of direct and tax subsidies, most of which have the effect of reducing the tax liability of energy producers or users.[8] (We set aside for the moment the federal government's role as the nation's largest energy user, and the large influence on innovation it can wield in some energy markets through its procurement activities.)

Federal subsidies of one form or another are in place currently for almost every type of energy technology and resource. Examples include production tax credits for coal-based synthetic fuels, wind, and other renewable sources; fuel blender tax credits for ethanol and biodiesel; tax credits for residential energy efficiency improvements; accelerated tax depreciation provisions for investments in transmission lines and nuclear plants; and preferential tax treatment of expenses incurred in oil and gas exploration and development. Of course, not all of these tax measures were designed to encourage energy innovation—if anything, many of them have had the opposite effect and have helped incumbents "lock in" their advantages in the market.

Table 2.1 summarizes federal energy subsidies by type of energy resource and type of subsidy. (Within each broad type of subsidy there are many individual programs.) The table invites many questions. Why the dominance of tax subsidies relative to direct spending on energy RD&D? Why this particular distribution of subsidies across different energy resources and technologies, and between electricity and non-electricity-related technologies? How effective are the different kinds of subsidies when it comes to achieving outcomes such as carbon dioxide emission reductions per dollar of public expenditure? Or reductions in energy imports per dollar? Or reductions in energy cost? Or jobs created? (Applying these sorts of outcome metrics to government spending on early-stage research is problematic since there are usually long time lags between the research activity and its eventual impact at scale and since so many other factors play a role along the way. But for government subsidies aimed at supporting innovations later in the innovation cycle, applying these kinds of metrics is more appropriate.)

The federal government has not provided an overall accounting of the impact of its energy subsidies. One program that has drawn particularly strong criticism is the decades-old system of subsidies for producing fuel ethanol from corn. Originally introduced as a way to reduce oil imports, the main beneficiaries of this program have been the agricultural sector, the fuel blenders, and the politicians who support these interests. But the

**Table 2.1**
Federal financial interventions and subsidies in energy markets, 2007 (millions of 2007 dollars)

| | | | | | | Electricity vs. non-electric | |
|---|---|---|---|---|---|---|---|
| Beneficiary | Direct expenditures | Tax expenditures | R&D | Federal electricity support | Total | Support for electricity production | Subsidies unrelated to electricity production |
| Coal | | 290 | 574 | 69 | 932 | 854 | 78 |
| Refined coal | | 2,370 | -- | -- | 2,370 | 2,156 | 214 |
| Natural gas and petroleum liquids | | 2,090 | 39 | 20 | 2,149 | 227 | 1,921 |
| Nuclear | | 199 | 922 | 146 | 1,267 | 1,267 | |
| Total renewables | 5 | 3,970 | 727 | 173 | 4,875 | 1,008 | |
| Biomass & ethanol/biofuels | | | | | | 36 | 3,249 |
| Geothermal | | | | | | 14 | 1 |
| Hydroelectric | | | | | | 174 | |
| Solar | | | | | | 14 | 184 |
| Wind | | | | | | 724 | |
| Landfill gas | | | | | | 8 | |
| Municipal solid waste | | | | | | 1 | |
| Unallocated renewables | | | | | | 37 | 360 |
| Hydrogen | | | | | | | 230 |
| Transmission and distribution | -- | 735 | 140 | 360 | 1,235 | 1,235 | |
| End use | 2,290 | 120 | 418 | | 2,828 | | |
| Conservation | 256 | 670 | | | 926 | | |
| Non-fuel-specific | | | | | | | 3,597 |
| Total | 2,550 | 10,444 | 2,819 | 767 | 16,581 | 6,747 | 9,834 |

*Source:* U.S. Energy Information Administration, Office of Coal, Nuclear, Electric, and Alternate Fuels, "Federal Financial Interventions and Subsidies in Energy Markets 2007," SR/CNEAF/2008-01, April 2008.

program (and associated tariffs on cheaper ethanol imports) has done little to reduce either energy imports or—despite claims to the contrary—net carbon dioxide emissions, and it has been quite costly to taxpayers and consumers. All of this has long been known, yet the program continues. But even leaving aside egregious examples like corn ethanol subsidies there is no reason to expect that the allocation of public resources described in table 2.1 is optimal with respect to twenty-first-century energy innovation goals. Each subsidy program was developed individually. Some were introduced many years or even decades ago, when policy priorities were quite different. Different agencies and often different Congressional oversight committees were involved. DOE, which might have provided an overall coordination function, has rarely done so—even for its own programs.

The Obama administration took office in 2009 with ambitious plans to accelerate the low-carbon energy transition. It quickly announced an increase in vehicle fuel economy standards as well as plans to establish greenhouse gas emissions limits for new cars and trucks. Rules to regulate carbon dioxide emissions from power plants were introduced as well. The legal basis for these new policies had been established two years earlier, in a 2007 U.S. Supreme Court decision that affirmed the authority of the Environmental Protection Agency (EPA) to regulate greenhouse gas emissions under the Clean Air Act.

In addition, the economic stimulus package introduced in the spring of 2009 directed an unprecedented level of resources toward energy-related projects—more than $80 billion altogether. As a result, for just one year DOE's energy budget ballooned from its previous level of about $5 billion to $39 billion, with much of the new funding targeted to energy efficiency and low-carbon energy supply technologies, including demonstration and deployment projects in domains such as utility-scale renewables, carbon capture and sequestration, weatherization, advanced battery manufacturing, and the smart grid.[9] Other federal agencies also received infusions of energy-related funding under the stimulus package. The General Services Administration, for example, received more than $5 billion for energy-efficient "green" buildings.

The Obama administration has been unable to put federal funding and policies for energy innovation on a more permanent footing, however. The proposed centerpiece of such a policy—a national cap-and-trade system for carbon dioxide emissions, supported by a variety of RD&D and other innovation programs—failed in the Senate in 2010. A shift in the political landscape after the Republican Party won control of the House of Representatives that November further undercut chances for climate legislation. Many new members of Congress profess deep skepticism about the reality of climate change, while others believe that the federal government either cannot afford to take aggressive action to reduce greenhouse gas emissions or cannot do so effectively even if money were available. Pressure has intensified to reduce federal spending in all areas, especially energy. Meanwhile, DOE's budget is falling back to prestimulus levels or lower.

Looking beyond the federal level, many states and localities have enacted policies to encourage energy innovation. Large chunks of stimulus funding were passed through to the states, for instance to support energy efficiency programs for buildings. But much state activity preceded the stimulus package, and some states and localities have gone well beyond current federal policies.[10] Twenty-nine states, for example, have adopted energy portfolio standards. These standards require utilities or other entities involved in energy delivery to secure a specified share of their supply from designated energy sources, such as renewables. The standards vary widely with respect to the quantitative targets they set, the time frame they provide for reaching targets, and the treatment of specific technologies (often in the form of "set-asides"). Most states have also adopted tax incentives, loan programs, rebates, or other supports for energy efficiency and low-carbon energy supply investments.

California, which comprises about one-eighth of the U.S. economy, has taken particularly aggressive actions to promote low-carbon energy options. California has long been well out in front of the federal government in the adoption of energy efficiency standards for buildings and appliances and fuel efficiency standards for automobiles. In April 2011 the state legislature affirmed its commitment to a renewable portfolio standard that requires California's electric utilities to have 33 percent of their retail sales derived from eligible renewable energy resources in 2020. A ballot proposition that would have suspended California's greenhouse gas emissions reduction policy was defeated in a statewide vote in 2010.

Regulation at the state and local level also plays a key role in affecting the pace of energy innovation, especially in the electric power sector, where state regulatory authority is pervasive. As we shall see in more detail in chapter 3, state regulatory commissions exercise a powerful influence on utility decision making with respect to the adoption of new technologies. The complex and fragmented structure of the electric power industry itself and of the regulatory regime that governs the industry at the state level has frequently been a hindrance to energy innovation.

Differences among state- and local-level energy policies reflect many influences. State and local economies vary widely in the ways that they use energy owing to differences in lifestyle, climate, population density, and industry composition. Their varied histories and geographical locations have bequeathed a diversity of energy resources and infrastructure, from the extensive hydropower complex in the Northwest to the nuclear plants of the Southeast. State and local political systems also reflect a range of views about the proper role of government, relationships between business and government, and the respective competencies of the public and private sectors. While fragmented governance is a problem in managing large energy infrastructures, such as the electric grid, it is also an asset in the sense that many different approaches to energy policy have been tried on a small scale in the "laboratories of democracy," as Louis Brandeis famously called the American states.

## Energy Innovation Policy: Beyond the Conventional Wisdom

The federal and state governments have had, and continue to have, an important and multifaceted role in energy innovation in America. Unfortunately, this influence has been at best uncoordinated and too often counterproductive. At the same time, it is clear that market forces are not sufficient on their own to bring about a large-scale transformation of today's energy system. So, the question is not whether government has a role to play in augmenting the market forces that drive innovation. Rather, the question is: what more should the government do, or what should it do differently, to unlock the extraordinary resources of the U.S. innovation system, especially in the private sector, so as to accelerate the energy transition?

Some leaders and commentators have called for a crash program by the federal government—a Manhattan Project or Apollo program for energy innovation. Such calls may be helpful as rallying cries, but not as blueprints for the task ahead. The energy challenge is different in almost every respect from those extraordinary initiatives. In Manhattan and Apollo there was a single, unambiguous, high-risk technical goal—a bomb that worked in the former case, a trip to the moon and back in the latter. There was only one "customer": the federal government itself. Success meant the technology had to work only a few times. In each case the goal was to be achieved in just a few years. And cost was essentially no object; the government was committed to doing the job, whatever it took.

None of this applies to the energy innovation problem. In this case there are multiple and sometimes conflicting goals: reduced carbon dioxide emissions but also affordable and reliable energy, not to mention increased energy security, good jobs, and less local environmental damage. There are also many different kinds of users. The federal government is of course a major customer—the Department of Defense by itself is America's single largest energy consumer—but most energy is consumed by a vast array of private organizations and individuals, from giant firms like Walmart and AT&T to every homeowner, tenant, and car-driver in the country. Innovations are needed on multiple time scales, from a few years to many decades. Success will come not from a few implementations; it will come only if low-carbon innovations are adopted by great numbers of users. And energy is a commodity, so cost competitiveness is vital. A one-time crash program will not do the trick.

A second simple answer to our question about the role of government that has wide currency is to get energy prices "right." The idea is that the government should estimate the total cost of greenhouse gas emissions and incorporate that cost into what users pay, just the way that prices for goods and services generally incorporate the cost of labor and materials that were required to produce and deliver them. A "carbon price" would induce consumers to reduce their use of energy and/or shift to lower-carbon energy sources. Economist William Nordhaus of Yale University puts the position bluntly: "To a first approximation, raising the price of carbon is a necessary and sufficient step for tackling global warming. The rest is at best rhetoric and may actually be harmful in inducing economic inefficiencies."[11]

Whether indirectly (through a cap-and-trade mechanism) or more directly (via a carbon tax), carbon pricing has the virtue of encouraging users to become more energy-efficient and to adopt low-carbon technologies that are available on the market at the time the price is imposed.[12]

The government need not be involved in selecting specific energy technologies or fuels to support, a process that is sometimes inefficient and at worst invites abuse or failure. Owners of the oldest, least-efficient, and most carbon-intensive equipment, such as World War II–vintage coal-fired power plants, would have the strongest incentive to shift their investment toward low-carbon alternatives.

While it is only possible to guess at the true cost of climate change in order to set the "right" carbon price, we agree that a price at some noticeable level ought to be part of any energy transition strategy. Prices are an important, probably the most important, element of what makes a market work. But a carbon price—though necessary—is far from sufficient to stimulate energy innovation adequately. For one thing, prices are not the only element of market functioning that must be addressed by energy innovation policy. For instance, buyers must have enough information to make good decisions. If they do not know how much they are paying for energy or what they might do to reduce their energy consumption, then simply raising prices will not be very effective at changing behavior. Government policies that mandate information disclosure or otherwise help market participants make sound decisions are complementary to policies that get prices "right."

Many economists would go along with policies that allow consumers to make more informed judgments. Many would also accept public funding of R&D. The argument here is that private investment is stymied by the "free rider" problem: businesses cannot prevent competitors from gaining access to the knowledge their R&D investments generate and cannot force competitors to pay for it either. Firms therefore tend to under-invest in these activities (relative to the amount of investment that would be justified if all benefits to society were taken into account), and public investment is required to compensate. This problem is especially acute for basic research, where the outcomes are the most uncertain and any commercially useful impacts are far in the future. Energy is no different from other sectors in this regard, and the solution in all cases is the same: direct and indirect financial support from the government for fundamental research and early-stage applied R&D, including grants to academic scientists and tax credits for private-sector R&D investments.

Our argument, however, goes beyond this conventional wisdom by focusing on barriers to energy innovation in the demonstration and early adoption phases. Demonstrating and scaling up complex energy systems; integrating them into an even more complex energy infrastructure; understanding and adapting technologies and business models to consumer behavior—these kinds of tasks are expensive, slow, and cannot be done in a laboratory or a simulation. Billions of dollars must be staked on these "learning investments"—in some cases, many billions. Most important, such investments will have to be made in the context of competition with fossil fuel–based energy systems that have been refined and optimized over a hundred or more years and that are supported by powerful constituencies with strong political relationships.

A very high carbon price would be needed to overcome the advantages that the incumbents have built up—a much higher price, in fact, than is necessary. The reason is that early in the learning process, the unit price of energy that an innovation can supply is often very high, precisely because there is so much that can still be learned about the innovation. This unit price is the one with which the incumbent competes. The gap between the innovator and incumbent may be so large at this point in the innovation process that a modest carbon price will not bridge it. If this is the case, the incumbent will continue to be the rational choice for most energy consumers. However, if learning investments can be made, the unit price of the innovation's energy output may decline over time. In fact, the innovator's costs may come down so far that the price gap with the incumbent will close to within the range of the modest carbon price referred to above, making the innovation attractive in the market. Ideally, the "learning curve" of the innovation will allow its price to dip below that of the incumbent. But without public support allowing innovators to discover whether a learning investment will pay off in this fashion—either that or an extremely high carbon price—this pathway is blocked.[13]

This argument inverts the Nordhaus view. Policies that effectively mobilize learning investments, allowing private innovators to bring down the cost of low-carbon energy, will in turn lower the carbon price that is needed to induce the energy transition. As long as innovation policies are less expensive cumulatively than the total cost of imposing a very high carbon price, they will reduce the economic impact of the transition on energy users and on the economy as a whole.[14] This strategy may have political benefits, too, since elected officials will presumably be more inclined to push forward with policies to accelerate the transition if those policies are less costly to their constituents. Whereas Nordhaus concentrates on increasing the cost of doing the wrong thing (emitting carbon dioxide), we call for parallel efforts to reduce the cost of doing the right thing (using low-carbon energy).

Although the learning investment argument is persuasive in theory, many practical objections can be lodged against it. Public decision-makers are less well informed than their private-sector counterparts about the business and market considerations that come to the fore in the middle stages of the innovation process. The government's administrative procedures are poorly adapted to market-based decision-making. Government involvement in demonstration and early adoption is more likely to be inefficient and politicized than its involvement in early-stage research, where peer review can be counted on to limit these problems.

As we have seen, such objections are well founded. The federal government has squandered a great deal of money on energy-related projects and programs that became white elephants, sinking billions of taxpayer dollars into some technologies even after their viability was clearly in doubt. Nor have state governments done much better. Some state renewable portfolio standards appear likely to be extraordinarily wasteful if they are fully implemented, requiring utility ratepayers to pick up the tab in future years. More likely, they will be modified or cut back as the true costs become evident, especially when the elected officials who first promoted them have left office and their successors realize that they have been stuck with the bill.[15]

If the stakes were not so high, we might well accept these objections and err on the side of the status quo. The threat of climate change changes the calculus, however. Now the risk that government intervention will fail outright or will be costly and inefficient must be weighed against the possibility that success will substantially lower the cost—and increase the feasibility—of the effort to accelerate the low-carbon energy transition. Concerns that the risk of ecological catastrophe will be used to rationalize even the most ill-advised adventures in government intervention must be balanced against the uniquely high ecological and economic risks of inaction given the nature of the climate threat. We are not saying that the practical objections to a much more active energy innovation policy should be ignored. They should be recognized, and policies designed that avoid the most serious pitfalls. But taking them too much to heart and erring on the side of the status quo would lock in a bad outcome.

In summary, on the key question of the appropriate role for government in support of energy innovation, this is roughly how the experts line up. There is strong support for macro- and microeconomic policies that would strengthen the general environment for innovation throughout the economy—for example, policies that promote a stable macroeconomic environment, well-functioning capital and labor markets, an effective regime for protecting intellectual property, and—crucially—improved educational performance at all levels. There is strong support for a price on carbon dioxide and other greenhouse gas emissions. And there is strong support for a direct government role in promoting basic research related to energy innovation. There is also some support for an additional government role in directly or indirectly encouraging the middle stages of the energy innovation process, but there is no consensus on how to do this, or even whether it can be done effectively at all. Some experts—especially economists—strongly believe that it is a mistake even to try.

We argue that it is essential to try. The stakes are too high not to. But equally important, the opportunities are also great. We find strong grounds for optimism that innovative technologies, new business models and new kinds of organizations can be demonstrated, refined and scaled up sufficiently between now and 2050 to mitigate the worst effects of climate change. The critics are right to stress the importance of learning from the mistakes of the past, though the successes should not be ignored either. It will be necessary, too, to think carefully about how energy innovation policies and government institutions interact with private entrepreneurs, with innovative users, and with large, established corporations—especially if the goal is to devise policies and programs that work not in isolation but rather by stimulating the American innovation system to develop in the energy sphere in the ways that have made it such a powerful source of innovation for the world in other domains.

These subjects are taken up in the following chapters. But the previous discussion has already begun to suggest some of the key requirements for building a successful energy innovation system. We summarize them here.

## Toward an Energy Innovation System that Works: Seven Principles

How to build an innovation system capable of sustaining an accelerated flow of energy innovations over a period of decades? From the perspective of the system as a whole, there are seven essential building blocks for success:

1. *The energy innovation system must support all four stages of the innovation process.* To accelerate the flow of energy innovations over a sustained period, all four stages of the innovation process must be stimulated. The two public policies that are most commonly advocated today to accelerate energy innovation are (1) to invest more in fundamental energy research and (2) to put a price on carbon emissions. Each of these policies will be necessary, but even in combination they will be insufficient. The impact of a carbon price would be felt throughout the innovation system, but would primarily accelerate improvements-in-use of technologies and business models that are already reasonably well developed. Increased government spending for fundamental research has the potential to accelerate option creation. But these two policies will have much less impact on the equally important middle stages, where many of the biggest obstacles to innovation arise. Much of the force that energy R&D spending and carbon pricing might exert at the beginning and end of the innovation process will simply be dissipated unless effective policies to address the middle stages are in place, too. Which of the four stages of the innovation process should be emphasized depends on the specific innovation. Figure 2.2 maps the four stages onto each of the three waves of innovation that were sketched out earlier. The focus is different for each wave. Today first-wave innovations are clustered toward the later stages of the process, second-wave innovations occupy the middle stages (demonstration and early adoption), and the third wave is predominantly in the option creation stage.

2. *The scale of the energy innovation system must match the scale of the challenge.* In the nearly forty years since the oil price shock of 1973 first brought energy issues to national prominence, energy innovation efforts have waxed and waned as economic conditions and political winds have changed. Transforming the energy sector will require a more stable, predictable, and much larger-scale effort than we have seen in recent decades. Going further back in U.S. history, one finds feats of innovation comparable to what this generation of Americans and its successors must undertake. The verdant forests of New England, which were stumpy, stony fields 150 years ago, attest to the shift from wood to coal during

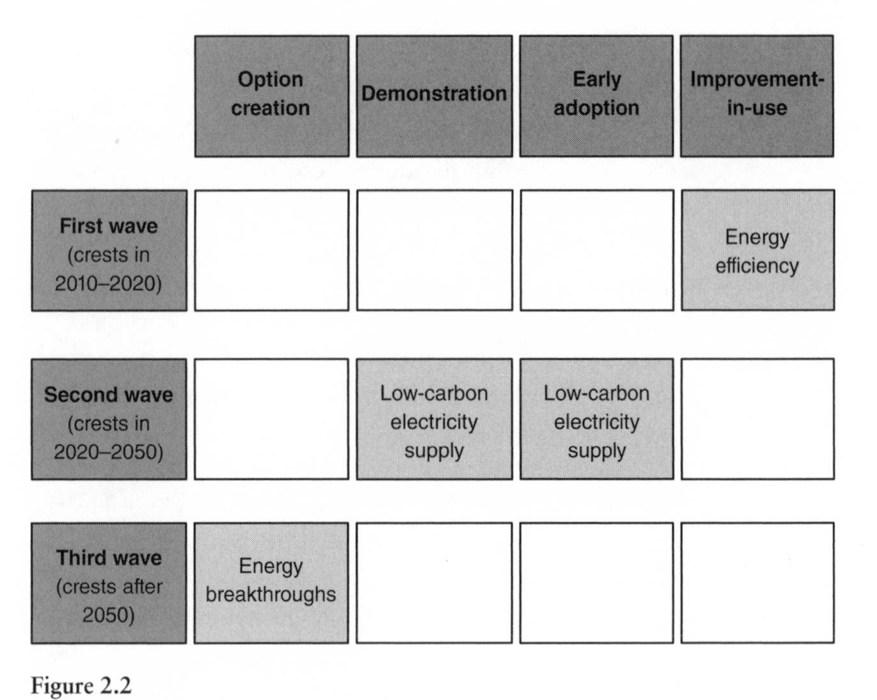


**Figure 2.2**
Energy innovation priorities today.

the nineteenth century. Under "normal" circumstances, the United States might naturally wean itself from fossil fuels by the end of the twenty-first century as the costs of extraction rise, cheaper alternatives slowly become available, and lifestyle changes demand new forms of energy services. But our circumstances today are not normal. Americans have just a few decades to bring about change on a scale that took their forebears a hundred years, in a society that is vastly larger and more complex than theirs.

3. *The great variety of ways in which energy is supplied and used calls for a correspondingly diverse innovation system.* Just as no single technological "silver bullet" will be found to address all the different needs of a carbon-constrained global economy, so too will there be no single organizational solution to the innovation problem. Tiny entrepreneurial startups, huge infrastructure providers, venture capitalists, pioneering user communities, and many others will be part of the mix. Consumer products like lighting technologies and energy-efficient window shades are developed and brought to market very differently from central-station

electricity generating systems like nuclear power and carbon capture and storage, and neither innovation process bears much resemblance to the way innovation occurs in the automobile sector. Accelerating the transition to a low-carbon, high-efficiency energy system will require a broad range of public roles and capabilities that reflect these variations in the innovation system. Some innovations will require little involvement on the part of government; others will require a lot.

4. *Competition between technologies, between organizations, and between business models must be encouraged at every stage of the innovation process.* Competition spurs effort and creativity. Markets are especially good at stimulating and resolving competition among innovators because they allow the users of a new product or service to decide what combination of features and price are most attractive, while sanctioning innovators who fail to respond to this feedback. Although this ability of markets to adjust to new information and to self-correct can never be fully replicated in the public sector or in hybrid public-private institutions, policies aimed at supporting innovation through the demonstration and (especially) the early adoption stages should take full advantage of markets and market-like mechanisms to drive improvements and inform winnowing down to the best options or applications. Beyond the early adoption stage, public policy should promote unrestricted competition between incumbents and innovators so as to maximize the pace of improvements-in-use.

5. *New entrants, including both new firms and existing firms from other sectors, must be encouraged.* The energy industry should remain in predominantly private hands. The private firm is a remarkably flexible and innovative organizational form, especially when subject to competition. No other kind of organization can scale an innovation as quickly. But the American energy industry needs an infusion of new firms, new people, and new ways of doing things. Public policy should create space for these new entrants and should also facilitate their access to resources.

6. *Timely winnowing at each stage of the innovation process is essential for limiting the cost of the energy transition.* Winnowing (or "down-selection") is important for both public and private investment. Feedback from users and other market participants is critical to these down-selection decisions, especially in the later stages of the innovation process.

Venture capitalists have developed systematic, rigorous methods for identifying the most promising options and for terminating the flow of funds to unpromising technologies and underperforming organizations. These methods can never eliminate all uncertainty, of course, and up to half of the investments in a typical venture capital portfolio either lose money or at best break even. There is no reason to expect public officials to do any better. Indeed, their task is intrinsically harder as they must consider not only economic viability but also the complex societal trade-offs that come with many energy innovations. Still, government decision-making for energy innovation has sometimes lacked discipline. Decisions have often been based on "point" design studies of a single technological configuration rather than using modern modeling and simulation techniques to select the optimal design from the full range of possibilities. Practical information about cost and operating performance has been ignored, too.[16] Public support for new technologies has sometimes been continued long after their unsuitability has become obvious, while in other cases promising options have been prematurely abandoned in order to conserve resources or to leave the field clear for the lead options. Ironically, one of the causes of these failures is the fear of failure itself, which often makes policymakers risk averse. Vigorous competition and systematic, rigorous down-selection will not avoid failure. The process of innovation always involves wrong turns and dead ends when technical barriers refuse to yield to expected solutions or when users turn out to have different preferences than anticipated. If the energy innovation system is to succeed, public officials will need to take risks. The most important objective is not to avoid failure—rather it is to ensure that failure is recognized, understood, and dealt with without delay.

7. *The energy innovation system should accommodate and exploit regional variations in innovation conditions and priorities.* Rapid progress on the low-carbon energy transition will require decisive action at the federal level. The fragmented structure of energy regulation, especially in the electric power sector, is a serious obstacle to innovation. Stronger federal authority will be critical in efforts to upgrade the nation's electric transmission system, implement more stringent energy efficiency standards for electrical appliances and buildings, and in other areas. But regions, states, and local jurisdictions also have essential roles to play in the nation's energy innovation system. Many critically important authorities—to make

land-use decisions, to set zoning requirements, to support public education, and to promote economic development, just to name a few—reside at the state and local levels. New industries will develop in support of the energy transition and state-level policies promoting the adoption of new technologies and the development of a skilled workforce can have a powerful beneficial influence on outcomes. Geographical proximity enhances interactions between research and educational institutions, investors, entrepreneurs, innovating larger firms and other key participants in the innovation process. Government policy cannot create these local innovation systems, but it can encourage their development. Traditional top-down federal approaches to funding energy RD&D at individual institutions should therefore be augmented by federal support for local and regional innovation systems. The partial regionalization of federal energy innovation policy will also enable innovators to exploit differences between regions of the country in terms of their resource endowments, economic and business conditions, and public attitudes toward energy options. Regionalization will also create new opportunities for inter-regional competition around important innovation outcomes such as technology adoption.

An energy innovation system that has these seven broad features will be much more likely to deliver the sustained flow of innovations needed to support a fundamental transformation than the innovation system we have in place today. But it is not enough to just to state these requirements. The question is how to achieve them in practice. That is the subject of the rest of this book.

# 3

# Electric Utilities and the Three Waves of Energy Innovation

The American energy innovation system of the future must generate many new options, test them, and weed out the least promising candidates on the basis of the test results. As innovations move from the option creation stage of the innovation process to improvements-in-use, the system should carry out these functions repeatedly on a wider and wider scale, but with a steadily narrower range of variations on the innovation at each stage. Over time, a speculative idea will become a highly optimized product or service with features that enable profitable business models to operate.

Markets and market-like structures play particularly important roles in the weeding-out process, especially in the later stages of innovation. The demand side of the market determines which features users want and how much they are willing to pay for them. The supply side creates the variations in technologies and business models from which users get to choose. Government organizations may be involved in some fashion on both sides of the market—and their involvement may be critical—but the most fundamental role of the government as a whole is to see to it that the market's institutional underpinnings are in place to the greatest extent possible.

Nowhere is this task more difficult or more important than in the electricity sector. A future low-carbon energy economy will rely more on electric power than does today's economy. But the electricity industry's unique properties present unique challenges for making markets work well. The power system is capital intensive, complex, and ubiquitous, and it has to respond instantaneously to changing conditions. These factors led the United States to limit the role of markets in the production and delivery of electricity after the underlying technologies required to operate the system reached maturity some eighty years ago.

The drawbacks of this approach became evident over time, prompting a partial "restructuring" of the electricity sector in the 1990s that aimed to introduce market forces more fully. But restructuring remains incomplete nationally and efforts to introduce further reforms have faltered, freezing in place a patchwork structure that is poorly suited to meet today's energy innovation challenges. In this chapter, we argue that completing the restructuring process can help to jump-start innovation in electricity generation, transmission, distribution, and use. Reformed and reinvented "smart integrator"[1] utilities will need to be central players across all three waves of innovation that we hope to see in the twenty-first century. But to unlock the full innovative capacities of the U.S. economy, these utilities will have to share the stage with the kinds of firms that they and their protectors in government crowded off in the past.

## Electricity: The Central Front

In chapter 1, we argued that the only way for the United States to achieve the ambitious carbon dioxide emissions reduction goal of "80 in 40" is to rely more heavily on low-carbon electric power than it does today. The point is worth revisiting in some detail here because it is too often overlooked or misunderstood. The electric power sector is *the* central front in the energy transition.

Emissions can certainly be reduced in the short run by improving efficiency, as we discuss in chapter 4. Switching from coal to natural gas can also cut power plant emissions for a time. In the transportation sector, next-generation biofuels may provide a means for cutting emissions as well (and would help the nation kick its petroleum habit to boot). Indeed, a transportation system based on carbon-neutral biofuels may even be achievable in the second half of the century. But between now and mid-century the United States faces a stark choice: slower economic growth *or* missing the "80 in 40" target . . . *or* transforming the electric power sector.

Simple arithmetic helps to illuminate these trade-offs. Let us take President Obama's target for 2050: an 83 percent reduction in annual U.S. greenhouse gas emissions compared with 2005. Staying with the assumptions that we made in chapter 1—an economy growing at a modest 2 percent per capita per year, a fairly aggressive 3 percent per year reduction in energy intensity, and population growth of 0.9 percent per year—carbon emissions per unit of energy used would still need to be cut 84 percent by 2050. So even a 50 percent reduction in emissions, such as might be provided by, say, replacing gasoline with second-generation cellulosic biofuels, is simply not enough. Compounding the challenge, there will be some activities or sectors of the economy for which no credible fossil fuel alternative exists. This means that wherever fossil fuels *can* be replaced by a zero-carbon option they should be. At present, such options exist only or primarily in the electricity sector. And that is why the only way to get to the "80 in 40" goal requires expanding the use of low-carbon electricity. There are no other large-scale low-carbon forms of energy on the 2050 time horizon. Genetically engineered liquid fuels that are carbon-neutral, for instance, have not yet been fully developed, much less deployed in a socially acceptable way. Passive solar technology for direct heating of air and water is an exception, but it proves the rule since it is confined to a limited range of applications and locations.[2]

The provisionally good news is that there are many low-carbon electricity technologies already in the field, although most of them still cost too much to compete directly with fossil fuel–based technologies. Generating systems based on hydropower, nuclear, wind, solar, and geothermal may be part of a sustainable mix. Coal and natural gas with carbon capture may be made to work as well. Moreover, electric power is extremely versatile: it can be used in a wide range of applications and it is highly transportable and convertible into other forms of energy. For example, electricity already competes with petroleum in some transportation applications; these applications must be expanded greatly in the coming decades, as we noted in chapter 1.

So electricity must be the dominant energy form used in the future. And as it turns out, the electrification of America's energy system is already an old story. The share of primary energy used to generate electricity in the United States has more than doubled over the past fifty years and now exceeds 40 percent. But almost 70 percent of the kilowatt-hours (kWh) of electricity that Americans consume today come from coal and natural gas (with no carbon capture). We must focus the creativity of the nation's innovation system on changing this pattern.

### From Frontier to Hodge-Podge

A century ago, when the electric power industry was young and dynamic, it harnessed the kind of creativity we are talking about. Electricity was an exciting new frontier for America's engineers, entrepreneurs, and managers. The names of many of the individuals involved are still well-known: Edison, Westinghouse, and even Vannevar Bush, who is better remembered today for contributions to information technology and science policy than for pioneering achievements in his home discipline of electrical engineering.[3]

To be sure, some cooling of this innovative spirit was inevitable. Technology matures, problems become more mundane, and the risks of departing from the tried-and-true mount, especially in a vital infrastructure sector. But in the power industry the slowing of innovation has gone much further than it might have. Its institutional structure has made it less an innovator than an obstacle to innovation today.

That institutional structure—which we define to include operating organizations, regulators, and even users—is quite complex. It can only be understood through the lens of history. In the United States, the institutional structure of the electric power sector is largely a by-product of event and circumstance, rather than a rational construction or an adaptation optimized to fit its environment. It embodies what social scientists call "path dependence," meaning that the structure in place today would be unlikely to be replicated even if it were possible to go back in time and start all over again.[4]

Like most young industries, the electricity industry of the late nineteenth and early twentieth centuries was a crazy-quilt of competing systems, business models, and governing ideas. By the 1920s, though, it had gone through a process of consolidation personified by Samuel Insull, the system builder who ran Commonwealth Edison and related power companies in the Midwest. After Franklin D. Roosevelt was elected president, his New Dealers went after Insull and other utility magnates, bringing them fully within the compass of the regulatory state.[5]

The idea at the root of the reforms introduced during the New Deal era (one that Insull also accepted) was that the electric power system is a natural monopoly from power plant to electrical outlet. Competition was considered to be wasteful in this context, since it leads to duplicative investments that cannot pay off. Firms that are granted monopoly power, however, must be regulated, so that they do not abuse their position by raising prices above the level required to earn a "fair" return on their investments.

The result by 1940 was an industry dominated by vertically integrated, investor-owned utilities (IOUs), most of which had relatively small service territories. Operating under the supervision of state public utility commissions, the companies were obligated to provide universal access to electricity with a high degree of reliability. In exchange for fulfilling that mandate, the IOUs were guaranteed a profit. Regulators fixed the price of electricity and limited entry into the market to ensure this outcome.

In addition to vertically integrated, state-regulated IOUs, the natural monopoly problem was also solved through public ownership. Municipal power companies ("munis") and rural cooperatives ("coops") popped up throughout the early twentieth century, especially in the Midwest and West. The New Deal brought to life two regional behemoths: the Tennessee Valley Authority (TVA) and the Bonneville Power Administration (BPA). These federal power authorities served multistate river basins where electricity service had previously been scarce. (See box 3.1, The Electric Power Industry—a Lexicon.)

How well this system worked during the ensuing decades is a matter of some dispute. In many respects it functioned well: demand grew, prices declined, and utility stocks generated stable returns for widows and orphans lucky enough to own a few shares. On the other hand, the system also gave rise to many costly inefficiencies, including excess generating capacity, obstacles to using the transmission network to access low-cost power, and many small utilities that were unable to take advantage of economies of scale.[6] Then the world changed. Energy prices became volatile, quadrupling in the 1970s and crashing in the 1980s. The rate of inflation rose rapidly, as did interest rates, and large, capital-intensive nuclear and coal-fired power plants experienced massive construction cost overruns. The environmental movement emerged, winning new pollution control requirements for existing power plants and fighting the industry over how and where to build new ones. In other sectors the natural monopoly concept was called into question from both right and left, yielding deregulation in the natural gas, railroad, airline, and telecommunications industries.[7]

In the electric power sector, these forces contributed to a partial dismantling of the New Deal system. As early as 1978, Congress created opportunities for small-scale generators to sell into the grid under the Public Utility Regulatory Policies Act of 1978 (PURPA). The 1992 Energy

**Box 3.1**
The Electric Power System—A Lexicon

The U.S. electric power system is complex and will likely become even more so in the future. Technologically, the system incorporates an enormous number and diversity of devices, stretching from the source of power to the user. The way the system is currently organized reflects the legacy of early entrepreneurship, later consolidation, and both successful and failed reform initiatives. It spans multiple levels of operation, from local to national. This box defines the terms we use to discuss the electricity sector in the main text.

*Generation* refers to the production of electricity from a primary energy source.

*Central-station generation* encompasses all large power plants, typically above 100 megawatts of electrical generating capacity (MWe).

*Distributed generation* is any form of generation that is not central station. Distributed generators are much smaller than central-station power plants—typically these systems are just large enough to power the facility, individual residence or office building, group of buildings, or small community where they are located.

*The grid* is the network of cables, switches, and control devices that carries electricity from the generator to the user.

*Transmission* refers to the high-voltage lines and associated infrastructure that transport power in bulk from central-station power plants to substations close to population centers. These are the "superhighways" of the electric system. Voltage on the transmission system is usually 110 kilovolts or higher.

*Distribution* refers to the medium- and low-voltage part of the grid that carries power from the transmission network to end users.

*Electric utilities* engage in some or all of the functions involved in supplying electricity to users, i.e., generation, transmission, and distribution.

*Vertically integrated utilities* have end-to-end ownership and control of all three of these functions. Utilities may be owned by private investors, municipalities, states, or the federal government—alternatively, they may be organized as cooperatives.

Policy Act established a broader framework for the operation of interstate wholesale electricity markets under the supervision of the Federal Energy Regulatory Commission (FERC). Many states followed suit by requiring utilities to separate the power generation side of their operations, in cases where competitive wholesale markets for electricity existed, from their transmission and distribution functions, which would remain

*Independent power producers* (*IPPs*) are firms that own and operate central-station generation and sell power into the grid, but that are not part of a vertically integrated utility. In general IPPs are not directly subject to economic regulation.

*Transmission companies* (*transcos*) are firms that own transmission facilities, but that are not part of a vertically integrated utility.

*Regional transmission organizations* and *independent system operators* (*RTOs/ISOs*) are nonprofit organizations that operate (but do not own) transmission facilities. RTOs/ISOs administer wholesale power markets.

*Federal power marketing authorities* are federal agencies that own generation and transmission facilities in the southeastern states (Tennessee Valley Authority) and the northwestern states (Bonneville Power Administration).

*Retail power companies* purchase power from the grid and sell it to end users.

In addition to the kinds of organizations listed above, an array of *publicly owned power companies* and *cooperatives* own and operate generation, transmission, and/or distribution facilities and sell power to end users at the local and regional level.

*Demand management* refers to measures that reduce system-wide power demand by limiting end uses of electricity at specific times. Some firms in restructured power markets specialize in organizing users and managing their demand. Utilities may also manage demand.

*Storage* makes it possible to use electricity that was generated at one point in time to meet demand at a later point. The only technology for storing electrical power on a large scale that has been commercialized to date is pumped hydro. This technology involves using electricity to pump water from a lower reservoir to a higher reservoir. The water can then be released at a later time to flow downhill and generate power. The ability to store electricity could be especially valuable in combination with intermittent generation resources such as wind or solar power. Several new large-scale electricity storage technologies are currently under development.

*Public Utility Commissions* (PUCs) are state agencies that regulate the electric power industry mainly by setting prices and limiting entry.

The *Federal Energy Regulatory Commission* (FERC) is a federal agency that regulates interstate commerce in electricity. FERC has jurisdiction over multistate wholesale power markets and transmission lines.

monopolies. This model also enabled states that so wished to authorize retail competition. Where they have done so, customers can in principle choose which service provider they want to buy power from; that provider then has the legal right to deliver electricity to the customer through the monopoly-controlled grid.

But electricity restructuring was not implemented everywhere in the United States and it was not always implemented well. California's approach, in particular, was badly flawed. Wholesale prices were deregulated there but not retail prices. In 2000–2001, weather and market conditions converged to create a crisis.[8] Rolling blackouts ensued and the state's largest electric utility, Pacific Gas & Electric, went bankrupt. The California crisis brought the national electricity restructuring movement to a crashing halt, despite the fact that many other states had had more successful experiences.

The institutional structure of the U.S. electric power sector today is, as a result, a "hodge-podge," as our colleague Mason Willrich labels it.[9] In roughly half the country—most of the South, much of the West, and parts of the Midwest—electricity customers are still served by state-regulated, vertically integrated IOUs. In another third of the states the institutional structure of the power sector has been revamped; these states have implemented wholesale and retail competition and their utilities have

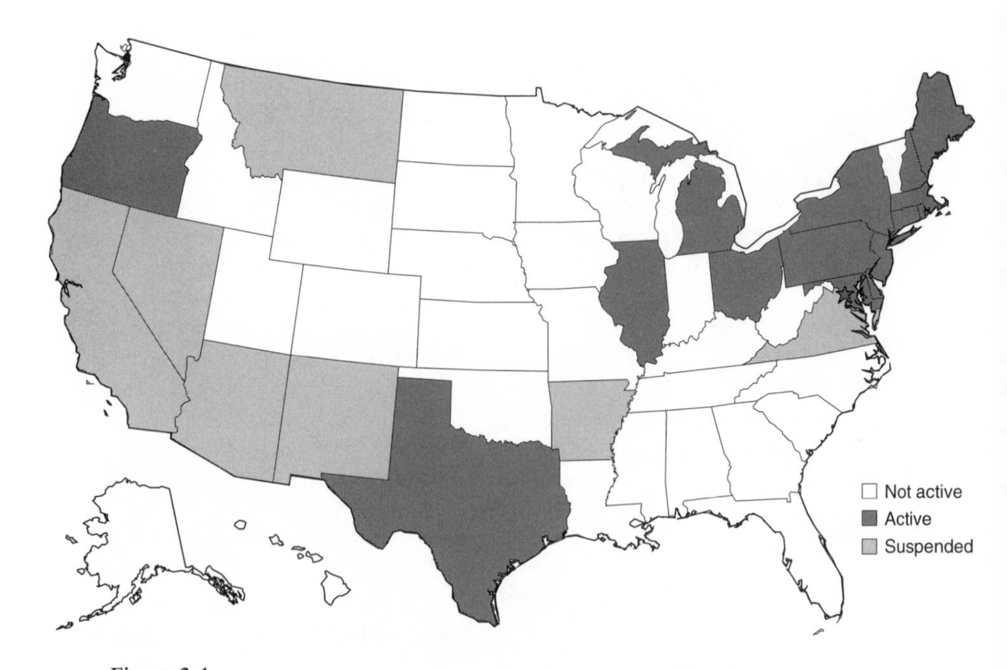


**Figure 3.1**
Electricity restructuring by state. *Source:* Energy Information Administration.

been restructured. The remaining states are stuck in a kind of limbo, with partially completed restructuring programs. Other institutional forms, including munis, coops, and federal power marketing authorities, are interspersed throughout the country (see figure 3.1).

## Opportunities and Obstacles

The relationship between this hodge-podge institutional structure and the innovation process is far from simple. Uncertainty about the future path of restructuring and regulation alone is a major disincentive to investment in all stages of the innovation process. The complexity of the current structure, with its fifty independent state regulatory commissions, its turf battles and border wars, complicates project planning and impedes risk-taking. While we acknowledge the virtues of institutional variety and regional diversity elsewhere in this book, they have become obstacles to innovation in the electricity sector. Indeed, our argument in this chapter is that the remnants of the New Deal institutional structure pose very significant barriers to innovation. That is true across all three waves of energy innovation that we have outlined.

- *Incentive Problems for Energy Efficiency* Greatly improved energy efficiency is the core objective of the first wave. In chapter 4, we focus particular attention on energy efficiency in buildings, which account for about 70 percent of overall electricity demand. Innovation in this domain has been impeded by the long reach of incumbent utilities, especially where vertically integrated IOUs still dominate. Their imperative to sell more electricity, in order to drive the revenue growth that investors look for, limits their interest in helping customers conserve energy. The concept of "decoupling" utility revenues from electricity sales is a relatively recent regulatory strategy that aims to remove the disincentives utilities face when it comes to supporting demand-side efficiency measures. But while decoupling can make utility revenues neutral with respect to the level of sales, experience with this approach is still relatively limited and it is unclear whether it offers a path to fully resolving the tension between maximizing revenues and minimizing energy use.
- *Large Investment Risks in Low-Carbon Electricity Supply* A key task of the second wave of innovation is to transition from today's

fossilfuel–based power generation systems, especially at central stations, to low-carbon technologies. This task is the main focus of chapter 5. Accomplishing it will require many large-scale, high-risk investments in first-of-a-kind plants. Even when the first of any kind of innovation is successful, the "next few" of that kind are also likely to carry significant risks. Uncertainties over future electricity demand, the price of natural gas (the most likely alternative fuel), and the price of carbon further affect the viability of such projects. Many utilities are simply too small to undertake high-cost projects and most of the rest are extremely reluctant to assume the risks involved. Where the New Deal regulatory system remains in place, regulators are often even more risk-averse than the utilities they regulate.

• *Accommodating Distributed Generation and Installing the Smart Grid* If the costs of distributed forms of electricity generation, such as small-scale solar and wind units, and of grid-scale electricity storage come down enough, these technologies may also play a critical role in the second wave of innovation, as we explore in chapter 6. Their viability will depend, in addition to lower costs, on the deployment of significantly improved transmission, distribution, backup, and control capabilities. These capabilities will need to be incorporated into the so-called smart grid. This term refers to an infusion of information and communication technologies into the electricity system that has the potential to match supply and demand more precisely and enhance the grid's overall efficiency, reliability, and resilience. This package of opportunities confronts many of the same risk and investment challenges as innovations in central-station generation. Moreover, distributed generation and the smart grid will demand the creation of innovative services and products on the customer side of the electrical meter, where traditional utilities have little expertise. In addition, new pricing models will be required to support these innovations—creating another enormous challenge to incumbent utility and regulatory practices.

• *Underinvestment in Long-Term Options* Even if the second wave of innovation comes to fruition in the coming decades, America in 2050 will face new and different energy challenges than it does today. The country should be exploring high-risk, long-range ideas that could form the basis for a third wave of energy innovation to unfold after 2050 (see chapter 7). However, neither the public sector nor the private sector supports or carries out very much of this kind of exploration. One indicator of the neglect is that IOUs spend only a tiny fraction of their revenues on R&D—the industry average is estimated to be below 0.2 percent[10]—a far cry from the 10 percent or so that firms in high-technology sectors, like pharmaceuticals and semiconductors, routinely spend.

In short, the electric power sector's old institutional structure generates too few options, has difficulty testing them, and often uses the wrong criteria to select among them. As later chapters will show, we would be the first to accept that restructuring, even when done properly, leaves gaps that should be filled by creative government policies. But we see the emergence of a vertically disintegrated electric power industry with a streamlined regulatory framework as a fundamental precondition for building a new energy innovation system.

## Toward the "Smart Integrator" Utility

Vertical disintegration has been a hallmark of change across many industrial sectors in recent decades, not least because it has allowed these sectors to more effectively seize opportunities for innovation. The "Fordist" firm (the term alludes to Henry Ford and the Ford Motor Company) that takes raw materials in through one door of an enormous factory and sends finished products out another, to be distributed and sold through its own retail network, financed by its own bank, and serviced by its own repair shops—this kind of firm is a dying breed. Yet the hierarchical structure of the Fordist firm has not generally been replaced by arms-length market transactions as some free market advocates have envisioned. Instead, supply-chain integrators or "lead firms" coordinate global supply chains, oversee product architecture and design, and integrate components into complex systems.[11]

While the electric power industry is unique in many respects, the basic pattern of change that we envision for it is similar in many respects to the pattern we have observed in, for example, the automobile and computer industries. Competition within horizontal segments—in this case wholesale power generation, retail electricity sales, and the provision of energy management services to end users—will become a stronger driver of innovation in the new structure than the centralized, top-down decision-making of the old vertically integrated structure. Nevertheless, some entity—the equivalent of a lead firm in other sectors—has to look out for the system as a whole after restructuring and ensure that innovation on the system's edges is compatible with its reliable functioning as well as its ability to meet carbon dioxide emissions reduction goals. That entity should be—must be—the "smart integrator" electric utility, operating under the watchful eyes of regulators and in cooperation with regional transmission organizations and independent system operators (RTOs and ISOs).

**Recommendation: Federal and state policy should aim to expand competition in wholesale electricity markets and in the provision of retail electricity and energy management services, while also giving utilities that have divested their generation assets incentives to focus on the critical task of system integration.**

"Wholesale electricity markets," writes the electric utility scholar Paul Joskow, "do not design themselves."[12] As painful as the California crisis was, it and other experiences in the United States and abroad over the past twenty years have demonstrated how these markets should—and should not—be designed. Independent power producers (IPPs) that own central-station generation facilities must have open and nondiscriminatory access to the grid, so that they can compete fairly against one another. RTOs/ISOs ensure such access, under rules set by FERC, at the level of the transmission system. IPPs bid to supply power based on locational marginal pricing, which reflects the availability of transmission as well as the attributes of the generator.[13]

From transmission systems that may be owned by a variety of different entities, but are administered by RTOs/ISOs, power in these restructured markets flows into the distribution systems that deliver electricity to end-use customers. Distribution systems are typically owned and operated by an incumbent utility that has sold off its generation assets. The distribution portion of the grid, too, must provide open and nondiscriminatory access, not only to central-station power flowing in from wholesale markets but also to power generated from smaller-scale distributed systems within the utility's territory and, in the future, perhaps from large-scale electricity storage facilities as well. The restructured utility serves retail customers of its own and interacts with competitive retailers who are using its distribution grid. It also deals with a variety of third-party service providers, including demand management aggregators (firms that work with groups of customers to reduce power demand at peak times). The ability to manage all of these diverse resources while also achieving system-wide goals for reliability, affordability, and emissions reduction under changing and sometimes unpredictable conditions is what will earn the restructured utility the label "smart."

The "smart integrator" path is not the only way forward. Peter Fox-Penner describes a second possible evolution of the U.S. utility business model, which he labels the "energy services utility." The energy services utility would be a new and improved version of today's vertically integrated electric utility. It would provide its customers with everything that they need that is energy-related. Its services might include expert advice, financing, software, and heating, lighting, and other end-use equipment along with kilowatt-hours of electricity. Fox-Penner points out that this model is a throwback to the early days of the industry, when Thomas Edison paid for and installed electric lights in buildings in order to expand the market for his electricity company.[14]

From an innovation perspective, the smart integrator model offers clear advantages over the energy services model. It is difficult to see how an energy services utility could create the conditions for rapid innovation. Success would require vertically integrated utilities to undergo a profound cultural transformation, shedding decades of conservative business practices while introducing innovative central-station generation technologies and diversifying simultaneously into innovative retail product and service businesses. Throughout, of course, energy services utilities would still have to fulfill their basic obligation to provide reliable, low-cost electricity.

The smart integrator business model would pose significant innovation challenges for utilities as well. The smart integrator utility would have to be agile, savvy about the use of information technology (IT), and sophisticated in its understanding of how the larger system will evolve over the long run. Smart integration also requires utilities to interact effectively with a great diversity of partners and rivals, both upstream and downstream, and with an equally diverse set of customers. The smart integrator would be responsible for creating an environment in which the most promising products and services thrive. Significantly, though, it would not have to devise or refine these products and services itself, nor scale them up. Those tasks would be performed by entrepreneurs and new entrants in competition with one another.

## Innovative Entrants in New Spaces: Independent Power Producers and Energy Management Service Providers

The withdrawal of utilities from layers of the electricity system that they have inhabited for decades under the New Deal framework, and the entry into these spaces of new players, is the strongest argument for the smart integrator model. Recent experience in parts of the country that have undergone restructuring provides evidence in support of our position. The remainder of this section focuses on the two largest and most important areas where new entrants and innovation are already playing a major role: wholesale power generation and energy management services for end users. (These and other parts of the industry are discussed in later chapters as well.)

First, however, we want to emphasize that vertical disintegration of the electric utility industry is likely to lead to unexpected ways of carving up the power system's functions, which in turn will enable innovative business models built around new bundles of products and services. These kinds of unforeseen niches have supplied much of the innovation that has emerged from other recently disintegrated industries.

IPPs now account for one third of electric power generation in the United States, a dramatic increase from just 3 percent a decade ago.[15] In that span, IPPs have made important innovations in fossil fuel plant technology, deploying new combined-cycle technology for natural gas units on a large scale, as well as fluidized bed combustion and selective catalytic reduction to control NOx (nitrogen oxides) emissions from coal units.[16] They have also led the way in dramatically improving nuclear power plant performance.[17] Finally, IPPs have dominated wind generation, which has accounted for the largest capacity additions in recent years. As of the end of 2009, IPPs owned 86 percent of cumulative installed wind capacity.[18]

These innovations are all the more noteworthy given that imperfections in the design of most American wholesale electricity markets appear to weaken incentives for IPPs to invest in generation facilities.[19] Part of the explanation can be found in a geographical circumstance: the parts of the country that have been subject to the most aggressive environmental regulation and that have enacted the most ambitious renewable portfolio standards are also the parts of the country that have carried out electric utility restructuring.[20] Nevertheless, it does appear that IPPs are more willing to take risks than vertically integrated utilities. In addition, IPPs may be willing to work more closely with new entrants among power plant equipment vendors, who are responsible for much of the innovation in this industry.

In theory, vertically integrated utilities might also have a large appetite for risk in the generation segment. After all, a fair return on their capital investment is supposed to be guaranteed by regulators. If regulators are willing to countenance risky investments in new power plant technologies, vertically integrated utilities ought to be willing to make them. In fact, this was the case in the 1960s and 1970s when nuclear power was introduced at scale in the United States. However, the American nuclear experience was characterized by long delays in construction projects and large construction cost overruns. These problems were exacerbated by, but were only partially attributable to, changing safety requirements. The legacy of this era includes not only public skepticism about nuclear power but also reluctance on the part of regulators to allow utilities to assume large-scale capital risk.

Another major area of opportunity for new entrants and innovation lies behind the meter on the premises of electricity customers. Households and businesses are not only using more electricity relative to other fuels, they are using it for more diverse applications. The proliferation of electronic devices is an obvious example. The eventual impact of battery-powered vehicles could be even more dramatic, possibly changing the overall timing and composition as well as the scale of electricity use in the United States. At the same time, these applications are becoming more flexible and capable as a result of advances in IT.

The growing diversification and flexibility of end-use devices are creating new opportunities to manage electricity use more wisely. Small commercial users and households are increasingly able to shift or limit demand in ways that were available only to large businesses and institutions in the past. Moreover, the control devices themselves have benefited from advances in IT, making them cheaper and allowing energy management to be automated more easily. These opportunities—the "customer-side smart grid"—have induced established IT firms and start-ups to compete with existing energy service companies in seeking business opportunities behind the meter.

Whereas the energy services utility would "own" the customer's energy needs, the smart integrator model stresses customer choice. Customers and the energy management service providers who serve them must have access to information about patterns of use in order to exercise this choice effectively. Incentives provided by time-varying pricing will also be critical to the effective operation of the smart integrator model. These issues are explored further in chapter 6. Our point here is simply that competitive third-party providers, along with customers themselves, are more likely to be innovative behind the meter—in the sense that they create and adopt new devices, software, and behaviors—than vertically integrated utilities of the sort recognizable to Thomas Edison.

In sum, the smart integrator and energy services utility business models imply two very different environments for innovation. The great weight of evidence on innovation performance from the electric power industry itself and from other sectors of the economy suggests that only shrinking the footprint of the vertically integrated utilities can create the conditions for a transformation of the power sector—the key to America's energy transition.

## The "Smart Integrator" in Context: Regulation and Coordination

The smart integrator, independent power producer, and third-party energy management services provider are all business models. Just as the vertically integrated, investor-owned utility business model was only viable under the New Deal regulatory framework, these twenty-first-century business models will succeed only if they are embedded in a supportive institutional context. The institutions surrounding these businesses, mainly public and quasi-public, must ensure that markets operate, and are perceived to operate, fairly. These institutions must also carry out functions for which smart integrators, IPPs, and energy management service providers are not well suited, especially with regard to innovation. In this section, we focus on two critical elements of the institutional structure: regulation and

RTOs/ISOs. We defer discussion of other institutional elements, such as our proposal for regional innovation investment boards, to later chapters.

**Recommendation: The analytical and deliberative capabilities of state regulators should be strengthened, so that they can more effectively design and enforce rules and incentives for smart integrators.**

Regulators, especially at the state level, will need to become smarter in tandem with the utilities they are regulating. The smart integrator utility will undertake a narrower range of activities than the vertically integrated utility, but those activities will become increasingly sophisticated as the number and diversity of resources being integrated by the utility grow. The smart integrator's continuing monopoly over the distribution system as well as its central position in the electricity information network will present many opportunities for it to act strategically.

This sort of strategic behavior by the smart integrator could well stymie innovation by IPPs, energy management service providers, and other third parties (even as it enables the smart integrator to take advantage of its own customers). The rationale for utility regulation will therefore remain strong even if the electricity system evolves in the way we are envisioning. Strict oversight will be required to maintain public confidence in these large, sophisticated, and strategically situated firms and to make sure they serve the public interest effectively.

An immediate challenge to regulators will be to establish rates that fairly compensate smart integrators for the system support and management services that they provide. Many of these services, like distribution and backup services, are inherently difficult to value. Some will be intangible and provide very diffuse benefits, such as assessing the system-wide value of innovative products and services offered by other companies. Since by definition no market exists to supply smart integration services, regulators will have to determine their value. To do so, regulators will need to develop more sophisticated analytical capabilities than they have had in the past.

Regulators and their legislative overseers will also have to make difficult judgments about tradeoffs between the goals of affordability, reliability, and greenhouse gas emissions reduction and between short-term and long-term considerations. We are particularly concerned about the tendency of regulators and legislators to focus on affordability in the short term, which sometimes leads them to seek low prices without regard to the potential consequences of these decisions for the system. The 2000–2001 California crisis provides a dismal example of where this tendency can lead. The closer the regulatory process is to the electoral process, the more pronounced the emphasis on keeping prices low in the short term is likely to be. Utility regulation has become much more politicized in the last two decades. Steps that would help insulate state regulators from immediate political pressure—for example, introducing staggered terms or requiring that candidates for certain appointments hold specific credentials—would be welcome. Recapturing and reinforcing the original concept of the independent expert regulator as an institution that employs outstanding people at all levels, rather than employing advocates for one interest group or another, are essential.

The enactment of a federal framework for reducing greenhouse gas emissions along with the continued development of a national electricity reliability policy would make things a little easier for state regulators, since it would allow them to blame Washington for requiring some costly investments. Stronger interstate and federal/state collaboration among regulators may also help them to do the right thing in the face of local opposition. At a minimum, such collaboration will help regulators share best practices and learn from mutual experience; moreover, collaboration will be essential to achieve integration and innovation at the regional level.

**Recommendation: Congress should extend the system of Regional Transmission Organizations/Independent System Operators to the entire country and grant RTOs and ISOs greater authority to plan and site new transmission lines.**

RTOs and ISOs are the nearest equivalent of a smart integrator at the regional level. In addition to operating wholesale markets and transmission systems that allow IPPs to sell and send power to downstream customers, they are responsible for ensuring reliability in compliance with mandatory standards set by the North American Electric Reliability Corporation (NERC). Most important for our purposes here, RTOs/ISOs carry out regional transmission planning within a regulatory framework set by FERC. Under FERC's framework, planning must be done in an open and transparent fashion that (a) provides ample opportunity for all stakeholders to be represented, (b) strives for consensus wherever possible, and (c) renders decisions on the basis of expert judgments.[21]

For smart integrators to provide retail suppliers and customers with access to new low-carbon generation resources that will come on line in the next few decades, new transmission lines will have to be built. Two recent studies commissioned by DOE's National Renewable Energy Laboratory (NREL), for instance, conclude that a major transmission build-out will be needed to allow wind power to supply a significant share of the nation's electricity.[22] The RTO/ISO process offers a proven means for sorting out these needs and opportunities. But at the moment these regional organizations are too weak to push through the actual siting of new lines, particularly when doing so requires taking on the entrenched power of state and local interests. As Mason Willrich states, regional planning decisions need to have "conclusive effect" in state and local proceedings.[23]

In those parts of the country where vertically integrated utilities or other entities rather than RTOs/ISOs take responsibility for the system's reliability,[24] such a planning process is not necessarily even available. The incumbent utility in these settings may use its power to block access to power resources from outside the area. FERC made an effort in the late 1990s and early 2000s to extend the RTO/ISO model to the entire country. But when electricity restructuring lost momentum at the state level, FERC's initiative petered out as well. Congressional action is needed to authorize and encourage renewed efforts to extend and rationalize the RTO/ISO system.

Many observers of the electricity industry, and in private moments even some of its leaders, wish that they could wave a magic wand and replace the fragmented regulatory structure, with its overlapping federal and state jurisdictions and fifty independent state regulatory commissions, with a single federal regulatory authority. A federal edict would be one way to achieve a more open industry architecture. But we do not advocate this. Any attempt to impose a top-down, centralized regulatory structure on the electric power sector would be time-consuming and enormously divisive, and would likely pre-empt most other reforms in the meantime. Instead, we urge an evolutionary approach, with stronger federal regulatory authority in some areas, more authority for regional groupings of states in others, and key roles for state and local authorities in still others. The details of this approach are laid out in later chapters.

Inevitably, some parts of the country will be reluctant to depart from the status quo. In some states, utilities and their regulators will opt for the energy services utility model. One or two of the most forward-looking states might even pull it off. But that cannot be the primary path to accelerated innovation across the industry. The better solution is to promote an open architecture, encourage open innovation, and create opportunities for new entrants to offer new services and products. If this objective is integrated into policy-making and regulation and the federal government and the states push consistently in this direction, the likelihood that a new, more innovative industry will emerge over time is greater than with a "big bang" approach.

## Conclusion

This chapter has made the case for an institutional restructuring of the electric power industry. The current institutional structure is, to put it bluntly, a mess. The remnants of the old structure and the unfinished work of restructuring combine to impede innovation today. Uncertainty about the future makes things worse. The situation threatens all three waves of innovation that are the main subject of this book.

To put things right, we recommend that electricity markets be extended and refined. We argue that market-based competition, an open industry architecture, and the entry of new competitors into newly opened segments of the electricity value chain will be a powerful driver of innovation. But these markets must be carefully designed and embedded in a supportive institutional framework, as recent history shows.

The key actor in our vision of the future is the smart integrator utility. The smart integrator will look out for the good of the system as a whole, managing the interaction of IPPs, distributed generators, energy management service providers, and other players. The smart integrator will address and resolve tradeoffs among equally important goals. It will have to be very smart indeed. Regulators, and RTOs/ISOs in their regional planning role, will have to nurture the smart integrator's intelligence and support its learning process across the entire nation.

If this restructuring can be achieved in the next decade, the United States has a fighting chance to achieve victory on the central front of the low-carbon energy transition. But to improve the odds for success the country will have to do much more.

# 4

# The First Wave of Innovation: Energy Efficiency in Buildings[1]

The transformation of the utility sector described in chapter 3 has already begun. It will take many years to complete, but other actions to accelerate low-carbon innovation can begin sooner. Indeed, waiting is not an option, given the scale and urgency of the challenge. Fortunately, there are many things that can be done right away to unlock the first wave of energy innovation.

The first wave is about getting more from less. People want energy services—transportation, lighting, heating, and the like—but each service could be provided with less energy than it takes now. The United States is the world's most profligate energy consumer. Without any technological breakthroughs and without even changing their behavior very much, Americans can cut their consumption of fossil fuels substantially in the coming decade while still getting the energy services they want.

That doesn't mean it will be easy. Many institutional and organizational barriers stand in the way of improving the nation's energy efficiency. It is in these realms—institutions and organizations—where innovation is needed most of all in the short run. This chapter concentrates on the biggest opportunities for first-wave innovation, which lie in the building sector. Our proposals would make it more likely that tenants and building owners will demand greater energy efficiency, and they would encourage a more diverse array of firms to satisfy this demand.

The innovations that we envision will yield a thriving marketplace for building energy efficiency products and services by the end of this decade. New entrants, new business models, and new mindsets will take root in this time frame. Over the longer term, in order to sustain the first wave into the 2020s and beyond, innovations in building design and technology will be needed as well, a point to which we return at the end of this chapter.

## Energy Efficiency: The Big Picture

Energy end use in the United States is divided among three big sectors. Industry uses about 30 percent of the total, while transportation, dominated by cars and trucks, accounts for another 29 percent. The remaining 41 percent goes into buildings, roughly half of it into homes, apartments, and other residential structures, and the other half into offices, shops, and other commercial structures.

The three sectors use energy in very different ways. Factories and refineries in the industrial sector use heat and electricity to melt steel, process chemicals, drive assembly lines, and so on. In general, these users have been more sensitive to energy prices than vehicle operators or tenants and building owners, because energy is a more important part of their cost structure. As a result, energy efficiency in industry has improved much more quickly than in either of the other two sectors in recent decades.[2]

Liquid fuels derived from petroleum power most vehicles. The fuel economy of the U.S. motor vehicle fleet has been relatively flat for more than twenty years. In 2010, however, new standards were established for cars and light trucks that will improve the efficiency of new vehicles about 25 percent by the 2016 model year. Tougher fuel economy standards for the years 2017 to 2025 were announced in August 2011.

Building occupants need energy to supply heat, hot water, air conditioning, and lighting and to power appliances. Most of the energy used in buildings comes from electricity, and building end-uses account for about 70 percent of the country's total electricity demand.[3] Electricity is also the fastest growing energy source for buildings. Given our heavy reliance on coal to generate electricity, it is no surprise that the building sector's share of carbon dioxide emissions is slightly greater than its share of energy use.

Progress in improving building energy efficiency has lagged behind the other sectors. Energy use per square foot in commercial buildings, for example, was basically flat from 1983 to 2003 (the most recent year for which data are available).[4] According to a senior official at the American Society for Heating, Refrigeration, and Air Conditioning Engineers (ASHRAE), a nonprofit organization that writes model energy codes for commercial buildings, the U.S. government "went to sleep"[5] in the building energy efficiency field after oil prices declined in the 1980s.

## Building Energy Efficiency: The Opportunity

Substantial opportunities exist to improve building energy efficiency. According to a recent National Research Council (NRC) report, the country could cut its energy use in buildings by 20 to 30 percent compared to "business as usual" over the next 20 to 25 years without any changes in technology.[6] Steve Selkowitz, who leads the building technologies group at Lawrence Berkeley National Laboratory (LBNL), put it this way in a 2009 workshop at MIT: "Building energy efficiency is not low-hanging fruit [for carbon dioxide emissions reduction], it is fruit that is lying on the ground rotting!"[7]

California, along with a few other states and cities, has bucked the national trend. Its success shows how much can be done right away. The state introduced a building energy performance code in the 1970s and, with technical assistance from ASHRAE and LBNL, has steadily tightened it since. The code has been updated, and made more demanding, every few years. For example, the California Energy Commission estimated that code changes between 2001 and 2005 alone would yield more than 20 percent in annual electricity savings in the state's low-rise residential buildings.[8] Between 2005 and 2008, the code's stringency increased an additional 15 percent.[9] Per capita electricity consumption in California is only about 60 percent of the national average, owing in no small part to more efficient buildings and appliances.[10]

International experience provides further supporting evidence. Germany, which has a harsher climate and an older building stock than California, has cut energy consumption per unit of floor area in new buildings by 70 percent since 1978, a figure that will rise to 80 percent when stricter requirements take effect in 2012. Retrofits of existing buildings that have been sold or renovated in Germany since 2002, when performance standards were put in place, have improved the energy efficiency of these buildings by about 30 percent.

Building energy efficiency is the most cost-effective climate change mitigation opportunity available to the United States. The NRC estimates that a *cumulative* investment of $440 billion over the next twenty years could produce *annual* savings of $170 billion in reduced energy costs. "The full deployment of cost-effective energy efficiency technologies in buildings alone," the NRC wrote, "could eliminate the need to build any new electricity-generating plants in the United States—except to address regional supply imbalances, replace obsolete power-generation assets, or substitute more environmentally benign electricity sources."[11] (The NRC cautions that this estimate assumes no "rebound" effect—that is, none of the energy savings are used to support increased consumption in other sectors.)

## Barriers to Improving Building Energy Efficiency

The barriers that prevent Americans from seizing this opportunity are almost as diverse as the building stock itself. There are schools and houses, office parks and apartment buildings, hospitals and hotels. Each type of building poses unique challenges. Climatic variation across the country, similarly, leads to a range of particular challenges in particular places: Santa Fe's issues are not the same as Saint Paul's. This diversity means that our review of the barriers to improving building energy efficiency will be thick with qualifiers and exceptions—nothing is simple in this sector! But this diversity of challenges also underscores the importance of experimentation and flexibility as policy-makers seek to address the barriers.

One important set of barriers arises from the fact that those who construct and own buildings do not necessarily pay their operating costs. Builders often seek to minimize their capital investment, leading to the construction of energy-inefficient structures. In doing so, they cater to buyers who are sensitive to the purchase price but less so to lifecycle costs. This is known as the "split incentive" problem because the builder has no incentive to make changes that will save the homeowner or tenant money later on. Rental properties suffer from similar split incentives. Building owners who would be burdened with the cost of efficiency improvements are disinclined to pay for them if their tenants will receive the benefits in the form of reduced utility bills. Within large organizations, there may be split incentives as well when one subunit pays the bill for efficiency upgrades while another receives the benefits.

Even where this issue does not apply, the initial cost of building energy efficiency improvements may be a barrier, particularly for retrofitting existing structures. High costs frequently dissuade homeowners from replacing inefficient building systems, such as windows and air conditioning units. For those willing to make the investment, access to financing may be difficult. Expected savings on future utility bills do not necessarily provide adequate assurance of repayment to potential lenders. Building energy investments can also take a long time to pay off. If the repayment period is longer than a lender is willing to accept or longer than the owner plans to own the building, the investment is unlikely to be made.

It is not surprising, then, that large, stable owner-occupants are the most likely entities in the building sector to invest in energy efficiency. Government agencies, universities, and large corporate headquarters units have the deep pockets and long time horizons that have been associated with such investments over the years. Institutions like these have been disproportionately represented, for instance, in the Leadership in Energy and Environmental Design (LEED) and Energy Star building rating systems, which reward energy efficiency. Such institutions are also better-positioned than other kinds of tenants and building owners to overcome information barriers that impede efficiency improvements.

Indeed, many who pay utility bills are ignorant about opportunities to reduce them. They rarely take the time to explore alternatives. Some tenants and homeowners may not have the expertise to analyze complex investments in equipment and paybacks over extended periods of time. They may expect investments to pay back their costs far more quickly than is realistic. In addition, the energy-wasting behaviors many consumers exhibit, such as leaving lights on and doors open, are habitual and hard to change.

A participant at an MIT workshop that we ran put it this way: "The product (i.e., energy efficiency) is so different. It's diffuse; it's invisible; it can't easily be measured."[12] Vendors of efficiency products and services have not been able to reach very many potential customers and inform them in an effective manner. Nor can utilities, many of which have incentives to maximize consumption, be counted on to do so, absent a push in the form of government policy.

Finally, there is the issue of prices. Building owners and occupants—even those who are fully informed and adequately incentivized—may find that the payoffs are just too small to justify the effort. A workshop participant provided the following example for a commercial building: personnel costs about $1000 per square foot; rent, $40 per square foot; and energy, $3 per square foot.[13] The value of a 1 percent improvement in the first category has over three hundred times more leverage for managers than a similar increment in the last category.

## Markets for Efficiency Products and Services

This diverse set of barriers to building efficiency improvements in the United States will not yield to a single "one-size-fits-all" solution. Establishing a carbon price would enhance the leverage of efficiency investments relative to others, but would not address the problem of split incentives. Technological innovations that lower the cost of making new buildings more efficient will not necessarily help owners of existing buildings.

The problem needs to be attacked from multiple angles at once. Innovations in financing can help to lower the initial cost of efficiency investments and reassure lenders of their value. Innovations in communication can make information about costs and payoffs more widely available and easier to understand. Innovations in market structure can induce competition that in turn spurs innovations in business models.

All of these innovations are aimed at invigorating markets for building energy efficiency products and services. The products, like low-emissivity windows and highly efficient air conditioners, already exist. So do the services, such as energy audits and shared savings plans for energy management. But they are currently confined to relatively small segments of the customer base. Stage 4 of the innovation process, where incremental improvement and widespread adoption go hand-in-hand, has not yet taken off.

Our proposals build on experimentation at all levels of government and across the private sector. While California is the acknowledged leader among the states, states as diverse as Utah, Hawaii, and Iowa are ranked in the top quartile in the 2010 energy efficiency policy rankings of the American Council for an Energy Efficient Economy (ACEEE).[14] Cities from Portland, Oregon, to San Antonio, Texas, to Arlington, Virginia, have undertaken building energy efficiency programs. Nonprofit organizations like the U.S. Green Building Council (USGBC) and for-profit businesses, like the USGBC's 16,000 members, are deeply engaged in the effort.

The federal government has been engaged as well. The General Services Administration, which manages buildings for most civilian federal agencies, adopted USGBC's LEED rating system for new construction in 2000, not long after it was first released. The Department of Defense, which manages about 2 billion square feet of building space, reported a

10 percent decrease in energy consumption per square foot for its facilities between fiscal years 2003 and 2009.[15] A series of legislative mandates and executive orders have pushed these agencies along.

Improving the energy efficiency of federal buildings is essential in light of the fact that the federal government is the largest single property owner and energy user in the country. But better management of the federal government's own activities is not an adequate substitute for a national building energy efficiency policy that reaches the vast majority of buildings, which are under private ownership. Such a policy would help to build national markets for building energy efficiency products and services that, in turn, would drive innovation on a national scale.

Some aspects of the policy that we propose in the following sections may seem to contradict our emphasis on markets and decentralization. We do call for tighter regulation of energy performance. We do suggest that the federal government play a stronger role in bringing about such regulation, although we would leave states and localities in charge of much of the regulatory process itself.

But this contradiction is only apparent. Regulation is not the mortal enemy of innovation, as some assert. Properly designed and with flexibility for states and localities to choose to go beyond a national baseline, regulation can and should be a central component in a system of innovation for building energy use and management. Such regulation induces entry, expands competition, enriches the choices available to building owners and tenants, and enhances the odds that they will avail themselves of these choices.

Regulation is most justified and most essential for new construction and major renovations.[16] These projects account for about 3 percent of the total building stock each year, which means that about 75 percent of the stock will be built or rebuilt by 2035.[17] Appliances, the fastest growing energy end use in buildings, would also benefit from tightened efficiency standards. The retrofitting of existing structures is less susceptible to direct regulation, but mandates that focus on the measurement and sharing of building energy performance information would make vital contributions to invigorating this market, too.

Of course, regulation is not the only tool in the policy maker's kit. Incentive and information programs are equally important. These measures promise to unlock the first wave of energy innovation.

## Energy Codes for New Buildings

The energy performance of new buildings in the United States is regulated unevenly, and enforcement is spottier still. The current regulatory system is cumbersome, with all three levels of government involved in implementing model building energy codes that have been drafted by nongovernmental organizations. Despite the risk of slippage in this system, which is made all too evident by the fact that energy-guzzling buildings continue to get built, we believe that the system's weaknesses are best rectified by tightening it up, rather than trying to replace it. One advantage of the current structure is that it embodies local/state/federal and public/private cooperation in a way that creates opportunities for experimentation, feedback, and continual improvement. But it is essential that a firm baseline be set, stuck to, and ratcheted up on a national basis.

**Recommendation: The federal government should provide strong incentives for states and localities to (1) put building energy codes in place for new construction where they do not now exist, (2) systematically strengthen those codes that are already in place, and (3) ensure that all codes are enforced. A collaborative effort to establish a nationally standardized building energy performance labeling system should support these codes.**

Building energy codes are descendants of building safety codes, which were born in the Progressive era of the early twentieth century. Fires, collapses, and other tragedies in the tenements of the country's burgeoning cities sparked reformers to enact safety codes, first at the municipal level and later at the state level. The energy crisis of the 1970s played a similar role in the development of building energy codes. The precursor of Title 24, California's code, was first authorized in 1974, for example. Some, but not all, states and localities followed California's lead in later years.

The role of the federal government in setting standards for building energy efficiency has been constrained by state and local prerogatives. In 1977 Congress asked the newly established DOE to develop national standards for buildings "designed to achieve the maximum practicable improvements in energy efficiency."[18] But when DOE tried to do so two years later, its proposals provoked such a fierce barrage of comments that Congress retracted the requirement, endorsing only voluntary performance standards.

DOE then enlisted ASHRAE and the predecessor organizations of today's International Code Council (ICC) to develop model energy codes for commercial and residential buildings. In 1992, Congress endorsed this arrangement, requiring states to adopt the latest standards promulgated by these organizations, as certified by DOE, or their own code of equal stringency. In recent years, the model codes have been strengthened considerably. ASHRAE's 2010 iteration of its commercial building standard (known as standard 90.1), for instance, is expected to be 25 percent more stringent than the 2007 version.[19]

A major weakness in the current system is that not all states comply with the law. As of May 2011 ten states had no commercial building energy code at all (although in many cases localities within them did[20]), and another fourteen had yet to adopt the latest certified version (ASHRAE 90.1-2007) or the equivalent. For residential building energy codes, the situation was slightly worse: 29 states had either no code or an old code in place. Frigid Minnesota and torrid Nevada, for instance, were one iteration behind in codes for both residential and commercial structures.[21]

Enforcing building energy codes is typically the responsibility of cities, towns, and counties, although the state also plays a role in many places. Even in California, the quality of enforcement varies widely from jurisdiction to jurisdiction.[22] Code enforcement is a labor-intensive activity, and budget and other priorities often impinge upon it. Layoffs of code inspectors as a result of the current recession have been reported around the country in recent months.[23]

These weaknesses might in principle be fixed by converting DOE's certification of model codes into a national regulation, thereby cutting out the step of state and local adoption. While this approach has the virtue of simplicity, proposals to federalize the building energy code will very likely arouse the same turf fights that vexed the 1970s attempt. Even if the federal government prevails, these fights might undermine state and local enthusiasm for enforcing the code. In addition, a single national standard could make it more difficult for forward-leaning states and localities to go beyond the minimum or to adapt building energy codes to their unique climate or building use demands. These "beyond code" efforts can play an important demonstration function, paving the way for future tightening of the baseline code.

Our suggestion is to strengthen rather than restructure the current system. Congress should give clear direction to the code-writing organizations and to DOE that each new iteration of any model code should ratchet up the efficiency goal by about 4 percent per year. The code adoption schedule might in the short run be negotiated with each state, to allow those that have been lagging a bit of leeway.[24] Failure to meet the agreed schedule, though, ought to be met with sanctions. For instance, funding for federally supported building projects might be withheld if stringent building energy codes are not adopted promptly.[25]

Enforcement, too, must be monitored and incentivized. Basic data on compliance with building codes have long been lacking. Federal/state cooperation to establish compliance rates is ongoing and must be strengthened. DOE's efforts to train state and local officials and to share best practices also warrant sustained support. Finally, Congress might consider financial subsidies for cash-strapped states and localities. Extrapolating from data gathered on code enforcement efforts in Massachusetts, we estimate that achieving 90 percent compliance might require about $1 billion per year to hire 16,000 inspectors nationwide.

A different approach to enforcement would be to rely more heavily on liability law. In Germany, architects and builders are legally responsible if code violations are discovered and are liable for 30 years if the building fails to achieve specified energy performance metrics. Large disparities between the energy performance predicted prior to construction and the actual energy consumption of the building can lead to investigations that uncover code violations. Our interviews with architects and builders in Germany confirmed that liability considerations drive adherence to code even in the absence of building inspections.

The success of tighter energy codes for new buildings does not depend, ultimately, on how ambitious they are nor does it depend on how aggressively they are enforced. Builders have to know how to build energy-efficient buildings. Buyers have to want them. With regard to the supply side, the evidence from California and elsewhere suggests that builders can and do adapt to tighter codes. A study for the city of Seattle, Washington, by the EDAW consulting firm in 2008 estimated that a 15 percent-20 percent increase in code stringency raised construction costs by only 1 percent-2 percent.[26] In Germany, building industry practices have been transformed. Builders have embraced that country's steadily tightening building energy codes and incentives for going "beyond code," creating a thriving market for energy efficiency products and services.

On the demand side, we believe that buyers, especially in the residential and small commercial markets, would be helped by a nationally standardized, easy-to-grasp building energy performance label and an associated rating system. Such a label would describe the building's expected operating performance and estimated costs over time. Labels of this sort have been found for many years on large appliances as well as on automobiles. Indeed, "MPG" is a common feature in auto advertising. The rating system used for Energy Star, a voluntary program administered by the EPA, can provide a starting point. At the moment, buildings in the Energy Star program earn a star if they exceed a design threshold, but more precise labels and ratings are not provided. New building labeling would complement the energy performance disclosure process for existing buildings that is discussed below.

"Green buildings" are a bright spot on the U.S. energy landscape. Enthusiasm for them is growing, and knowledge about them is expanding. A more consistent and predictably improving regulatory system, which engages all levels of government as well as professional and industry associations, will make the market for new buildings the foundation of the first wave.

### Appliance Efficiency Standards

The energy-efficiency regulatory process for new appliances and equipment is much simpler than that for new buildings. DOE sets standards and manufacturers implement them. The process has been proven to work well. Even the most hard-nosed economic analyses suggest that the standards have paid for themselves in consumer savings, without even accounting for their environmental benefits.[27] The problem has been inconsistent effort from DOE over time. The Obama administration has once again restarted the standard-setting process for appliances, and it is important that future administrations follow suit.

**Recommendation: The current drive to tighten federal energy efficiency standards for major appliances should be sustained.**

The most important energy services used in buildings, both residential and commercial, are heating, ventilation, and air conditioning (HVAC), lighting, water heating, and powering electronic equipment, including computers. Demand for some of these services has grown rapidly in recent years. For instance, the share of households with central air conditioning more than doubled between 1980 and 2005.[28] Electronic equipment was uncommon before 1980, even in offices. When efficiency improves, the use of these kinds of devices can grow without a comparable expansion in energy consumption.

Like building energy codes, appliance energy efficiency standards first appeared after the oil shocks of the 1970s. States like California, Massachusetts, and New York led the way. But unlike building construction, appliance manufacturing is dominated by a small number of large companies. These companies did not want to have to comply with many different state regulations and so supported an upward shift of jurisdiction over appliance standards to the federal level. Federal/state regulatory turf battles, not surprisingly, are far less pronounced in appliances than they are in buildings.

Thirteen household appliances were included in the first batch of federal standards, which were legislated in 1978. Refrigerators are the best-known success story from this early group. The average refrigerator used 71 percent less electricity in 2007 than in 1977, an annual rate of improvement of about 4 percent, even though the average 2007 model was larger, cheaper, and had more features.[29] The trends are the same, if less dramatic, across other product categories subject to energy efficiency regulations, and they have been validated by many studies.[30]

Congress has expanded the number of appliances covered by standards over time, including most recently in the 2007 Energy Independence and Security Act (EISA). The standards are supported by a variety of other policies as well as by voluntary programs. For instance, the federal EnergyGuide and Energy Star programs provide labels and branding to inform consumers about appliance energy efficiency. Federal tax incentives have helped create a demand-pull for greater efficiency since 2005 as well.

While Congress has shown resolve, even impatience, to strengthen appliance efficiency standards, the executive branch has not always followed through. President Reagan, who was first elected not long after the 1978 enabling legislation, was philosophically opposed to such governmental interventions in the market place. Litigation overturned this position but not before nearly a decade had passed. More recently, the George W. Bush administration dragged its feet. The Government Accountability Office reported in 2007 that DOE had missed all 34 deadlines for establishing new standards set by Congress.[31]

The Obama administration has made a significant commitment to clearing the backlog and accelerating the standard-setting process. The American Council for an Energy Efficient Economy reports that the promised pace of work "far exceeds what DOE has done at any other time in its history."[32] While the marginal gains to be had in some product categories may be diminishing as a result of existing standards, previously unregulated products offer significant energy and economic savings. ACEEE estimates that simply replacing fluorescent lamps in the commercial sector would produce more than 25 billion kilowatt-hours of electricity savings in 2020, or about 0.7 percent of current U.S. electricity consumption.[33]

When appliance standards work in concert with "beyond code" policies to spur efficiency improvements, standards can be ratcheted up steadily where that is appropriate. The efficiency levels set by the voluntary Energy Star labeling program for residential clothes washers in 2001, for instance, were adopted as a standard in 2007.[34] Similarly, a voluntary agreement in August 2010 between manufacturers and energy efficiency advocates established goals for future appliance standards that are based on levels of efficiency that previously earned federal tax credits.[35]

The only real barrier to continued energy efficiency improvements in appliances is federal willpower. Industrial concentration means that suppliers adapt to changes in the playing field. Customers, and the nation, benefit—sometimes without even knowing it.

## Incentives to Retrofit Existing Buildings

**Recommendation: Federal, state, and local governments should expand and simplify financial incentives for energy efficiency retrofits with the goal of attracting more private investment into the retrofit market.**

Existing buildings present far more difficult challenges than new buildings or appliances. The cost and disruption of installing energy-efficiency retrofits can be considerable, and often involve more than the buyer bargained for when purchasing the property. It is simply not feasible or fair to slice through the Gordian Knot of barriers with a regulatory mandate for retrofits. However, with an existing building stock of 129 million homes and nearly 5 million commercial facilities nationwide, and very low turnover in the building stock from year to year, the retrofit challenge cannot be ignored. The institutional structure for retrofits needs to be rebuilt so that it expands the market, strengthens competition, and enhances the information available to everyone involved.

**Expanding the Retrofit Market**

Building retrofits may involve anything from plugging leaks and beefing up insulation to installing and commissioning new HVAC systems.[36] Simple projects may require only information and a little time. More complex projects demand professional design and installation and the investment of a substantial fraction of the building's value in these services along with materials and equipment. Especially for owners of residential and small commercial buildings, sticker shock often makes retrofitting a nonstarter.

The most straightforward solution to this challenge is to give building owners money to help defray the up-front cost. Rebate programs do so directly, while tax incentives do so indirectly. Many states have put these kinds of policies in place, and the federal government reinstituted tax incentives for some energy-efficiency investments in 2005 after letting them lapse a couple of decades earlier. Such policies can work well, especially if they are well targeted to particular populations and applications, supported by education and stakeholder involvement, and sustained long enough for the market to popularize them.[37] However, financial incentive programs are often confusing, owing to their sheer diversity and frequent shifts in coverage. Some are simply wasteful. Careful design, better coordination, greater consistency, and simplification across the levels of government and across programs within each level would make them more effective.

Societal benefits charges (SBC) imposed by state public utility commissions are currently one of the biggest sources of funding for building retrofit programs. In many cases, SBCs were introduced as part of the electric utility restructuring wave of the 1990s. Most states now have programs of this type, and they are growing rapidly. Efficiency programs funded by SBCs (or similar ratepayer charges) on electricity use were budgeted at about $5.4 billion in 2010, twice the total for 2007. About $1.7

billion of this funding was spent on residential retrofit measures such as compact fluorescent light bulbs, air-conditioner tune-ups, and appliance upgrades, and on low-income programs.[38]

Ratepayer-funded energy efficiency programs have generally proven to be cost-effective. The energy that they have saved has been estimated to cost between 2 and 5 cents per kilowatt hour, much less than the average electricity retail price of about 10 cents per kilowatt hour—a good (albeit involuntary) deal for ratepayers.[39] But state-level SBCs are unlikely ever to fund more than a fraction of the total retrofit requirement. These programs typically pick up about half the cost of the improvement, with customers covering the rest.[40] If we assume an average retrofit cost per residence of, say, $7500, and a goal of reaching 3 percent of America's 129 million homes each year, the annual capital requirement would be roughly $30 billion. If half of this amount were paid for by an SBC, it would add more than a penny per kilowatt hour to household electricity bills. While ratepayers in some states are evidently willing to pay a societal benefits charge in the 0.1–0.3 cents per kilowatt hour range today, many would likely balk at a charge that is roughly five times greater.

These calculations suggest that SBC-funded programs will not scale adequately if they are designed primarily as rebates or other direct subsidies to building owners. The same problem applies to tax credits: when uptake of such credits begins to take off, the keepers of the public purse tend to pull back. Public dollars go further when leveraged by the capital markets, so that building owners mainly get loans instead of rebates. Using SBC funds to leverage private investment would be a more scalable financing strategy for residential and small commercial building retrofits than allocating these funds directly. The proposed federal Clean Energy Deployment Administration (CEDA) and similar proposals to create public "green banks" provide other means of achieving the same goal.[41] Public financial institutions abroad, such as Germany's KfW, have proven effective in this role (see box 4.1 on Germany's "Green Bank").

Innovations in servicing as well as sourcing loans will also help to unlock the residential and small commercial building retrofit market. One such innovation is property tax–associated financing (commonly known as property assessed clean energy or PACE).[42] Property tax–associated financing addresses a major barrier to retrofits by encouraging building owners to make investments that might not pay off until after they sell

**Box 4.1**
KfW—Germany's "Green Bank"*

Germany's KfW (Kreditanstalt für Wiederaufbau) banking group was founded in 1948 to promote reconstruction and economic development after World War II. Jointly owned by the German states and the federal government, the bank today raises most of its funds at competitive rates in the regular capital markets. As a nonprofit bank that is not required to pay taxes, its lending costs remain low. Interest rates are additionally subsidized through federal tax funds, and are generally about two percentage points below commercial rates. The bank is not allowed to compete with private banks, but rather cooperates with the commercial banking sector in providing affordable financing for energy efficiency retrofits and $CO_2$ reduction measures. The bank also provides a wide range of other services, including student loans, small business lending, and export financing.

KfW first began offering preferential loans for energy efficiency in 1996. For new buildings, KfW today offers preferential loans for the purchase or construction of buildings that use no more than 70 percent of the maximum energy consumption mandated by the current building code. Buildings with less than 55 percent of the maximum allowable consumption qualify for additional interest rate reductions.

KfW also offers low-interest loans for the purchase of existing buildings if they have been retrofitted to meet current code requirements for new buildings; further interest rate reductions are available if the retrofits go beyond these requirements. Additionally, KfW provides loans for energy efficient renovations. Loans for retrofits are subsidized to a greater extent than those for new buildings.

KfW's loans to individual applicants are processed by regular commercial banks, which advertise, perform credit checks, and collect payments in return for a fee. The banks are also required to verify that the loans are used for the intended purpose and that required efficiency levels have been reached. Since the size of the preferential KfW loans is capped, customers often rely on additional commercial loans to finance the remaining cost of the construction or renovation project. The banks thus usually offer customers a mix of commercial and public preferential loans, relying on the preferential loans to reduce the average interest rate.

According to the German federal government, €31 billion in KfW loans were originated in the four-year period through 2010, resulting in a total of €61 billion in energy efficiency investments in new and existing buildings over this period. This was achieved with taxpayer subsidies to KfW of just €1.5 billion per year. Thus, every euro in tax subsidies to reduce interest rates generated about €9 of investment in energy efficient buildings. Thanks to KfW's preferential loans and other incentive programs, the rate of energy efficiency retrofits as a percentage of Germany's total building stock now exceeds 2 percent per year, still shy of the government's 3 percent target, but many times greater than the rate in the United States.

*Excerpted from Jonas Nahm, "Energy Efficiency in Buildings: The Case of Germany," Draft Working Paper, MIT Industrial Performance Center, December 2010.

their properties. States and localities may serve as financial intermediaries under this approach, using their unique borrowing status to offer loans at submarket rates; funds from federal taxpayers, ratepayers, or commercial institutions might also be used. The innovation is that repayment of these loans is made through the property tax billing system. When a property changes hands, the buyer assumes responsibility for the energy-efficiency loan along with the property tax. Unfortunately, federal mortgage lenders and regulators (Fannie Mae, Freddie Mac, and the Federal Housing Finance Agency) have blocked the widespread diffusion of this innovation, because the property tax–associated debt is senior to the mortgage debt that they are financing. This conflict should be resolved by Congress, so that the hiatus in property tax-associated financing is ended as soon as possible.

On-bill finance is another innovation in loan servicing that promises to accelerate energy-efficiency retrofits for smaller buildings. Customers using this service repay loans for energy-efficiency improvements in existing structures through the same process that they pay their electricity (or natural gas or heating oil) bills. On-bill finance is obviously convenient, but in addition, it may lower financing costs because defaults on utility bills (as on property taxes) are relatively rare. One obstacle to on-bill finance arises when the utility is not the lender. A customer might choose to pay only for his energy costs and not the loan installment. The utility in this circumstance has little incentive to go after the delinquent borrower. Indeed, the utility may need to be compensated simply to use its billing infrastructure for this purpose at all. Supplying such billing services fits easily within our vision of the smart integrator utility.

## Strengthening Competition and Enhancing Information in the Retrofit Market

The administration of state energy efficiency retrofit programs varies from place to place. Utilities take the leading role in some states, such as California, while state government agencies and nonprofit organizations are the central players elsewhere. While all three approaches have achieved successes in the past,[43] our preference is for nonprofit administration. The core competence of smart integrator utilities lies in managing electricity transmission and distribution and in integrating diverse resources onto the grid. Efficiency will be one of many "plug and play" applications that these utilities will need to integrate.[44] State government administration is vulnerable to shifts in the political winds, as governors and legislatures with different priorities come and go.

Nonprofit administration of public or ratepayer funds for energy-efficiency loans and rebates does not mean nonprofit delivery of products and services. Rather, the administrators of such programs should be market makers who establish and enforce quality standards while fostering competition and innovation among providers, whether for-profit or nonprofit. This competition may occur among lenders as well as energy auditors, equipment providers, installers, and the like. Business model competition might yield new combinations of products and services, such as one-stop shops that include financing.

Program administrators can further support the retrofit market by making energy consumption data more easily available. At the moment, much of this information is locked up in utility IT systems.[45] Even with due attention to privacy protection—a necessary requirement to win public trust—such information should facilitate marketing efforts by third-party providers while also informing consumer choice. Legislation may be needed to give program administrators more authority to require disclosure.

The sale of an existing building also provides a critical opportunity to enhance consumer knowledge about energy efficiency. While a new building may be labeled, based on design expectations, in the same way that a fuel economy MPG label appears on a new car, actual performance data should be supplied to prospective buyers of an existing building. Disclosure of these data, converted into the same sort of standardized lifecycle performance metric required for new building labels, should be mandated in advance of any sale. An energy audit, perhaps bundled with the standard building inspection, should be required in advance as well, so that behavioral influences on the performance data (like the number of occupants and their habits) can be considered by the buyers. These audit requirements might be incorporated into building energy codes.

Creative retrofit programs are essential policy tools for jumpstarting the energy transition. But their contributions will quite literally be wasted if building users leave windows open in the winter and lights on when they leave home. Automation and intelligence built into energy systems can help to reduce mistakes and nudge users into better decisions, but innovation must extend to personal behavior as well as technologies, institutions, and business models. Building labels, better marketing, and other innovations should reduce the level of ignorance, but there will still be an important residual role for public information and education programs run by a trusted source. Energy-efficiency program administrators should be one such source. The ultimate goal of information and education is to instill a social norm of energy efficient behavior that reflects the true social value of such behavior.

If properly designed and governed, dedicated nonprofit energy-efficiency organizations at the state, regional, and local levels can become valuable repositories of expertise on program design and communication. The governing boards of these organizations should include representatives of a broad range of stakeholders, including utilities, building owners, and technical experts. The federal government should continue to help benchmark, provide technical assistance to, and foster information exchange among organizations that implement energy-efficiency retrofit programs; the government may also need to play a role in ensuring that effective accountability structures for these organizations are in place.

## Sustaining the First Wave with Design and Technology

The institutional innovations that we have recommended in this chapter can deliver a decade or more of substantial progress in building energy efficiency. "Low-hanging" fruit—and certainly any fruit that is lying on the ground—should be picked in these years. The gaps between the United States on the one hand and northern Europe and Japan on the other, and between the rest of the country and California, can be closed. At the same time, social norms around energy use may change such that efficiency gains greater acceptance as a priority for the average consumer.

Even with these changes, further headway must be gained to achieve the ambitious mid-century greenhouse gas emissions reduction targets outlined in chapter 1. To reach higher-hanging fruit in the 2020s and beyond, innovations in building design and technological systems will be essential. The basis for these advances must be laid in this decade through investments in research, development, demonstration, and diffusion.

One premise of this chapter is that there is a gap between the building technology frontier and common practice. This gap is mainly the result of practitioners' sluggish adoption of innovations—it is not as if the innovation frontier is moving so fast that users cannot keep up. The key reason innovation is so slow is that the building industry is so fragmented. Only about a dozen or so construction firms in the United States have more than 1,000 employees and fewer than 900 have more than 250 employees.[46] Few firms have the incentive or the resources to invest in R&D. The result is that "buildings," as a leading researcher once put it, "are the largest handmade objects in the economy."[47]

It falls to the federal government to play a central role in coordinating the creation and diffusion of knowledge in the building sector, much as it has in agriculture. The U.S. Department of Agriculture took on the mission of advancing agricultural science and technology in the late nineteenth century. The productivity of American farmers grew steadily for many decades as a result.[48] The Department of Energy has begun to play an analogous role in the building sector only recently.

DOE has created productive building energy efficiency teams at its national laboratories, including particularly Lawrence Berkeley National Laboratory and the Pacific Northwest National Laboratory. Arthur Rosenfeld, a key figure in the California efficiency success story, guesstimates that building performance modeling software, which represents just one output of the DOE labs, saves the country about $10 billion every year.[49]

Under the current administration, DOE's building technology program has set aggressive goals, such as creating marketable, net-zero energy commercial buildings by 2025. Commercial buildings, especially large ones, are very complex, so achieving this goal will require not only R&D but also the same kind of real-world systems integration and learning-by-doing that occurs in power plant demonstration projects (see chapter 5). DOE has created institutions that have the potential to carry out such projects and to disseminate the knowledge that they generate. These institutions include the public-private partnerships of its Building America program for residential structures and the brand-new "innovation hub" for energy-efficient buildings in Philadelphia.

A recent assessment by the American Physical Society called for sharply increased spending on next-generation building technologies and for training building scientists, but also warned against continuing the tendency to allow short-term activities to crowd out the pursuit of longer-term possibilities in DOE's building technology program.[50] More social science research on consumer behavior, including research on how consumers process information, will be an essential component of a longer-term program.

Beyond DOE, the rest of the federal government is poised to adopt new building designs and technologies quickly. The 2007 EISA requires that new federal buildings achieve the ambitious target of net-zero energy use by 2030.[51] The defense establishment has declared even higher ambitions. The Department of the Navy's drive to build a "Great Green Fleet," for instance, calls for at least 50 percent of its installations to be net-zero energy users by 2020.[52]

Buildings that are certified to a high level by nongovernmental green building rating systems, such as LEED, also provide opportunities for demonstration and early adoption of energy-efficiency innovations that go well beyond building energy codes. The availability of a wide range of public and private test beds should reinforce the continual ratcheting-up of these codes. By refining and debugging innovative buildings, such projects allow code writers to be confident that the requirements they impose with each new code update are technically and economically feasible.

The institutional structure for generating innovative energy-efficient building designs and technological systems, unlike the institutional structure for widespread adoption, appears to be relatively robust. It is primed to contribute significantly to the achievement of greenhouse gas emissions reductions, if it receives an infusion of resources and if these resources are sustained over the long run.

## Conclusion

Accelerating improvement in energy efficiency is the first wave of innovation that America needs to effect a transition to a low-carbon energy system. Building energy efficiency is the most important focus of this wave. In this chapter, we have sketched a restructured system of innovation that will speed the widespread adoption of building technologies that are already commercially available and push forward the technological frontier over time.

Our recommendations aim to invigorate the market for building energy efficiency products and services. For new buildings and for appliances, regulation that is strong and growing stronger in a predictable fashion will be the cornerstone for making these markets work better. For existing buildings, a combination of financial incentives, new financing institutions, new administrative structures, and new business models will be necessary. Better and more accessible information, such as an energy-use label for all buildings (similar to MPG labels for cars), will support deeper markets for efficiency products and services.

To repeat the central point of this chapter: the United States can cut fossil fuel consumption substantially in the coming decade while still getting the energy services that consumers and the economy demand. The products and services that would make this feasible already exist. So does a system for making many of these products and services even better. What has been missing has been willpower and resources.

The first wave of efficiency innovations should crest in the coming decade but it must not slacken after that. The wheel of progress in efficiency must keep turning if the targets suggested by our analysis in chapter 1 are to be fulfilled. But it is the second wave—a remaking of the electricity system—that will carry the greater load a decade and more hence. We turn our attention to that subject next.

# 5

# The Second Wave of Innovation—Part I: Low-Carbon Electricity Supply

Unlocking energy efficiency—not only in buildings, which we have discussed in detail in chapter 4, but also in transportation and industry—will be the first wave of low-carbon energy innovation. However, as we showed in chapter 1, the nation's energy and climate goals cannot be achieved through efficiency improvements alone. America's need for energy is too great, and the existing energy system too carbon-intensive, for an efficiency-only approach to work. Nor is natural gas-fired electric power a permanent solution, even though it emits substantially less carbon dioxide per kilowatt-hour than coal. While welcome at the moment, an excessive focus on gas could be unhelpful over the longer term if it deters investment in even lower-carbon alternatives. As a recent MIT study on the future of natural gas concluded, "though gas is frequently touted as a 'bridge' to the future, continuing effort is needed to prepare for that future, lest the gift of greater domestic gas resources turns out to be a bridge with no landing point on the far bank."[1]

In the medium and long term we must unlock a second wave of innovation that enables low-carbon technologies to supplant natural gas and other fossil fuels for power generation. A central goal of this effort must be to improve the performance and lower the unit costs of electricity production from low-carbon generation technologies whose basic scientific and engineering characteristics are already well known. The options include solar photovoltaic and solar thermal technologies, wind (including offshore wind installations), advanced nuclear reactor and fuel cycle technologies, carbon capture and storage technologies for both coal and gas-fired power plants, geothermal, and biomass. Large-scale deployment of second-wave innovations that prove to be viable should begin about a decade from now and continue for several decades.

Much has been said about which electricity generation technologies ought to dominate in the low-carbon economy. We do not take a position on the eventual winners and losers, but instead envision an experimental approach to determining outcomes. Indeed, the terms "winner" and "loser" are inappropriate, since the likely outcome will be different mixes of options in different locations and for different uses. The nation's portfolio of low-carbon generation alternatives is likely to include "all of the above."

We include in this assessment distributed generation as well as central station generation. Almost all electricity consumed in the United States today is produced in central-station plants, mainly coal (45%), natural gas (23%), nuclear (20%), and hydro (7%). Historically, the economies of scale and cost efficiencies afforded by central station generation have far outweighed the drawbacks, which include large up-front costs and vulnerability to adverse events. Innovations on the horizon, which involve pricing, distribution, and other aspects of the electricity system as well as generation, may increase the role of distributed generation. But the traditional dominance of central station generation will not fade quickly—indeed, it could persist indefinitely. (In chapter 6 we take up other aspects of second-wave innovation, including storage and the smart grid, and deal with the question of distributed generation in greater depth.)

A major institutional restructuring will be necessary to unlock second-wave innovations for electricity generation, whether central station or distributed. In particular, funding for the demonstration and early deployment of these innovations must be mobilized and allocated in new ways. As we argue in the next two sections, neither the market (including venture capitalists, some of the most creative participants in the market) nor the federal government has effectively supported these two stages of the innovation process in the energy realm. We then briefly review existing proposals to remedy this problem. The centerpiece of the chapter is our own proposal for a regionally based mechanism to fund demonstration projects and early deployment programs. Our proposal calls for upending established practices and injecting competition at multiple levels, building on the principles that we articulated earlier in this book.

## Gaps in Financing Demonstration and Early Deployment

An innovation system that will hasten the transition to low-carbon energy must support all four stages of the innovation process: option creation, demonstration, early adoption, and improvement-in-use. One of its most important jobs is to ensure the availability of adequate funds. As a rough rule of thumb, the funding requirement increases tenfold when an innovation moves from one stage to the next (see figure 2.1). A well-functioning innovation system will also attend to the "gates" between the stages, winnowing out options that perform poorly. The order-of-magnitude jump in cost at each gate puts a premium on doing this job well.

While the cost of option creation will presumably be covered mainly by public funding (see chapter 7), this stage is the least expensive of the four. At the other end of the process, the vast majority of funding for improvements-in-use will come from companies who are already producing the technologies that are being improved. The stages in between, demonstration and early deployment, logically require a mix of public and private funding. Just what the mix should be, and what rights the public and private partners ought to have with regard to winnowing out the options being evaluated, are hotly debated topics in both public policy and private investment circles. These issues have been so challenging to resolve that the passage through the two middle stages of the innovation process is sometimes referred to as the "valley of death."[2]

The problem is difficult even for relatively small-scale innovations, such as software and electronics for managing energy use. But it is much harder for large-scale central station electricity generating technologies. When the budget of a single demonstration project (or what we will refer to as the "next few" post-demonstration projects) runs to hundreds of millions or billions of dollars, the valley of death looks more like a chasm. Costs for the mass production of distributed generation technologies (such as building manufacturing plants for solar modules) during the early adoption stage may also be in this range. (Solyndra, a thin-film solar photovoltaic manufacturing company, had to raise $970 million in equity finance, in addition to receiving $535 million in federal loan guarantees, even before it was floated on the stock market in 2010.[3]) Yet, unless these early costs are underwritten by somebody, the energy industry will not learn how to make low-carbon technologies affordably and operate them reliably.

The sheer size of the learning investment is a deterrent in both stages, but there are additional barriers specific to each stage. At the demonstration stage, private investors are deterred by a large array of uncertainties that make it hard to assess the odds of success. There are uncertainties not only about the technical and economic performance of the technology being demonstrated, but also about safety and environmental regulations, the future market price of competing fuels and the regulatory price that may be levied on carbon dioxide emissions. Another deterrent to private investors is the difficulty of appropriating the benefits of investments in demonstration projects. One of the main purposes of such projects is to disseminate credible information about the cost, safety, and reliability of the new technology as widely as possible to prospective providers, investors, and users. For private investors, this requirement reduces the value of proprietary knowledge or intellectual property that such projects produce.

The learning that takes place at the early adoption stage, by contrast, should be proprietary, occurring within the firms that are building and operating projects and among the firms in the supply chain. The chief goal in this stage is to drive down costs. Nonetheless, the process of doing so may be lengthy and expensive, especially for the "next few" projects built after the initial demonstration has shown a technology to be technically feasible. Innovators in this stage must compete head-to-head against well-entrenched competitors who have spent many years honing their technologies. The cost of power generated by the "next few" will often still be above that of the incumbents. The cost gap may persist for some time. (Indeed, in some cases it may never be fully closed. Some technologies will simply fail the affordability test and will need to be winnowed out as a result.)

Some observers have suggested that venture capital can bridge the "valley of death" in energy innovation as it has in many other industries since the mid-1970s. Venture capitalists are accustomed to taking technical risks, and they routinely build learning curves into their investment plans. In fact, "clean tech" venture investments rose rapidly during the past decade, from $230 million in 2002 to more than $4 billion in 2008.[4] But most of this investment has been, by necessity, on a relatively small scale. A typical individual investment by such a fund rarely exceeds $10–$15 million. As a result, most "clean tech" investment has gone into distributed generation and smart-grid technologies, rather than large-scale central station generation technologies.

A second challenge for the venture capital model is exit. As the costs rise, venture investors will want to partner with or sell out to deeper-pocketed but more cautious investors. Yet the risk-return profile of investments in energy innovations in the early adoption stage is often unattractive for these "downstream" investors. This is because expected returns to innovation in commodity markets like electricity are fairly modest, but the risks and uncertainties are diverse and large. So downstream investors have often remained on the sidelines of large-scale clean tech, cutting off what has been a favorite exit route for venture investors in other high-technology industries. So far, for example, there have been few equivalents in the energy industry to the roles played by large pharmaceutical companies and information and communication companies as sources of scale-up capital. (More details can be found in box 5.1 on "Why Venture Capital Plays Only a Limited Role in Energy Innovation.")

**Federal Support for Demonstration and Early Deployment: The Record**

The failure of the market to adequately fund the demonstration and early deployment of low-carbon energy innovations is widely recognized. The federal government has frequently stepped in to try to fill these financing gaps in the past. But as we began to see in chapter 2, many of these attempts have failed. Troubled projects and perverse incentives have been all too common.

The history of demonstration projects managed by DOE and its predecessors spans several decades. An early, and by many measures successful, federal investment was the Shippingport project of the 1950s, which demonstrated the use of pressurized-water nuclear reactor technology for commercial-scale power production (the technology was originally developed for naval propulsion).[5] More recent projects have tended to be less successful. The Clinch River Breeder Reactor Project, for instance, cost U.S. taxpayers about $1.7 billion and never operated. The synthetic fuels demonstration program failed in its stated goal of producing a half million barrels a day of synthetic oil and gas from domestic sources, largely because the cost of its products far exceeded the market price of oil. The Yucca Mountain nuclear waste repository project has been mired in controversy since it was initiated in the late 1980s and is currently on hold. The saga of FutureGen, DOE's erstwhile flagship project to demonstrate carbon capture and sequestration, is still unfolding, but seems to be headed down the same troubled path.

**Box 5.1**
Why Venture Capital Plays Only a Limited Role in Energy Innovation

The basic model of venture capital investing entails the formation of a fund managed by a small number of general partners and financed by third-party investors or "limited" partners (limited in both liability and management control). There is no upper limit on the size of these funds, but the partnership structure and the ability of individual general partners to manage no more than a certain number of active investments means that funds typically do not exceed a few hundred million dollars.

Most early-stage companies with innovative technologies or business models will never become profitable, but venture capitalists are willing to invest in these businesses anyway, despite their high technological and market risks, because they expect a small fraction to generate high returns in the end. A typical individual investment made by such a fund therefore rarely exceeds $10–$15 million. If the amount were much bigger, the fund would not be able to make enough investments to achieve reasonable risk diversification. This structure constrains venture capital to less capital-intensive businesses. Venture capitalists also prefer to invest in businesses whose commercial viability can be established in five years or less, allowing them to exit well before the life of the fund—typically 10 years—is over. This approach enables them to establish the record of successful performance needed to raise the next fund.

Exit is achieved through the public or private equity markets, or by selling the business to an established company in the sector. Downstream investors typically augment their capital with bank loans or other kinds of debt finance. Their access to these capital markets provides them with much more capital than venture capitalists have at their disposal, but it also means they have a much smaller appetite for risk.

An ideal venture investment—that is, one that enables rapid and profitable exit—is one where the technological risks, while large at the outset, can be resolved fully within a few years and with only a modest amount of capital. Many software-based businesses have these features, with Google and Facebook being among the most spectacular recent examples. Some innovative energy technologies and businesses also have these characteristics, but many more do not. Large central-station generating technologies like nuclear reactors clearly do not. Nor do many solar, wind, and storage technologies, where uncertainties over regulatory requirements and energy pricing policies as well as important technological and economic performance risks cannot be resolved in advance of utility-scale deployment or full-scale manufacturing, or both. These residual risks are likely to be too great for risk-averse downstream investors to assume, while the capital requirements for scale-up are too large for individual venture funds or even for syndicates of funds.

Federal post-demonstration subsidies have also been problematic. Rather than stimulating innovators to bring down the cost of new technologies as quickly as possible, they have often had the opposite effect. Open-ended government subsidies have rewarded firms not for innovating but simply for producing regardless of cost. Probably the most egregious example is the now decades-old program of corn ethanol subsidies, but there have been other cases where an idea that seemed promising at the outset spawned an entrenched constituency that later resisted all attempts to scale back support. The federal government has typically been unable to ratchet down subsidies in order to drive cost reductions, much less shut projects and programs down in a timely fashion when they have clearly failed to produce the expected results.

Why has the federal government's track record in energy technology commercialization been so unsatisfactory? Excessive congressional involvement is one reason. Influential elected officials have sometimes interfered in technology selection and facility siting decisions and personnel appointments. Congressional pressure has also limited the ability of executive branch officials to adjust or terminate projects after conditions have changed. In other cases, Congress has forced the abandonment of projects even when conditions have not changed. The annual budgeting and appropriations process has invariably been a source of uncertainty for program managers, while policy shifts following congressional and presidential elections have added further to the lack of stability in funding and policies over the life of projects.

Other explanations focus on problems within the executive branch. Federal agencies including DOE have at times displayed a tendency to underestimate project costs (perhaps as a requirement to generate political support).[6] Restrictive federal procurement regulations and bureaucratic rules governing personnel decisions, auditing requirements, and the use of federal facilities have also been implicated in cost overruns and poor management. MIT's John Deutch notes that Congress has not granted DOE and other agencies the authority to attract individuals with the experience and skills to oversee major projects.[7] DOE and its overseers on the appropriations committees have not been able to establish and adhere to multiyear budget planning.

## Reform Proposals on the Table

The dilemma is evident. On the one hand, the prospects for successfully commercializing low-carbon electricity generation technologies, especially large-scale central-station technologies, will be very dim if the market is left to do this on its own. Some public role in sharing the costs and risks of demonstration projects, and in devising strategies for bridging the post-demonstration cost gap, seems essential. Yet it is difficult to avoid the conclusion that DOE is structurally unable to play this role effectively.

Several proposals have been advanced to break the impasse. Senator Jeff Bingaman, chairman of the Senate Energy and Natural Resources Committee, and ranking member Senator Lisa Murkowski have recently introduced legislation to create a new federal financing entity, the Clean Energy Deployment Administration (CEDA), that would give high-risk energy demonstration projects and deployment programs access to various forms of financing, including loans and loan guarantees. CEDA would be a semi-independent unit within DOE.[8]

Another proposal would go a bit further, creating a Green Bank as an independent, tax-exempt corporation that is wholly owned by the federal government. The Green Bank would be similar in structure to the Export-Import Bank. It would support diverse technologies and projects through debt financing and credit enhancement, giving priority to those projects that would contribute most effectively to reducing greenhouse gas emissions and oil imports. The Green Bank would be an independent, but government-owned, tax-exempt corporation with both public officials and experienced private individuals on its board.[9]

A third proposal, advanced by MIT's Deutch, would establish an autonomous, quasi-public organization, the Energy Technology Corporation, specifically to finance and execute large-scale energy demonstration projects.[10] It would have flexible hiring authority and follow commercial practices in its contracting. This corporation would be governed by an independent board of directors nominated by the president and confirmed by the Senate.

Finally, the American Energy Innovation Council, a group of leading business executives including General Electric CEO Jeff Immelt and Microsoft chairman Bill Gates, have proposed a public-private partnership to address the problem.[11] Noting that America's energy innovation system "lacks a mechanism to turn large-scale ideas or prototypes into commercial-scale facilities," this group has called for the formation of an independent, federally chartered corporation, outside the federal government, that would be tasked with demonstrating new, large-scale energy technologies at commercial scale. This new organization would "offer project management services and technical resources to help accelerate and improve the design and construction of facilities. . . . It would work to enable fast-track siting and construction opportunities within utilities or public power agencies, on federal or military lands, or even overseas . . . in some cases." It would report to a politically neutral, congressionally mandated Energy Strategy Board that would also be external to the federal government.

Though the details vary, all of these proposals are designed to overcome the limitations of DOE and to insulate innovation decisions from political forces to some degree. The new entities that have been proposed would be free of many of the most burdensome federal rules; they would also have much more flexibility in management and would be independent of the annual congressional budget cycle. But all of them would require a big one-time congressional appropriation to get started: $10 billion for CEDA; $10 billion for the Green Bank; $60 billion for the Energy Technology Corporation; and $20 billion for the public-private energy demonstration corporation proposed by the American Energy Innovation Council.

### A Regionally Based Demonstration and Early Deployment Funding Mechanism

We agree with the thrust of these proposals, though we do not think they go far enough. It will of course be quite difficult to secure a congressional appropriation for tens of billions of dollars in today's budget environment. More important, these proposals do not adequately embody the foundational principles for a durable and productive energy innovation system that we articulated in chapter 2. The most important of these principles include: creating space for new entrants (both new firms and existing firms from other industries), providing a reasonably steady flow of investment capital to innovators over long periods, ensuring timely and rigorous down-selection at each stage of the innovation process, accommodating regional differences, and matching the size of the challenge. In the remaining sections of this chapter, we put forth our alternative.[12]

A starting point for this exposition is EPRI, formerly known as the Electric Power Research Institute, a nonprofit R&D organization serving the electric utility industry. EPRI was established by electric utilities in 1972 after a series of massive power blackouts prompted concerns about the industry's performance. It is a membership organization with broad representation within the electric utility sector. Over the years EPRI's research agenda has been set by its professional staff in conjunction with their technical counterparts in utility companies. The research agenda is approved by the organization's board of directors, who are elected by the members. EPRI has made important research contributions although it has shrunk considerably since utility contributions to support its work were curtailed in the wake of the restructuring of the 1990s.[13]

The key features of EPRI from our perspective are its membership base and professional staff. EPRI members are the users of its work: the organization provides them with a unique opportunity to articulate their collective needs and informs them of technical trends. These features distinguish EPRI from the demonstration and early deployment financing models we described above. But EPRI also has weaknesses, including a lack of guaranteed funding and a near-monopoly position in its field. Moreover, the electric power industry has not pushed the innovation frontier with the vigor the climate challenge demands.

The scheme we propose has a set of EPRI-like membership organizations at its core. We call them regional innovation investment boards (RIIBs). As the name suggests, we envision creating several of these boards around the country, rather than having only a single one. The RIIBs would decide which large-scale demonstration projects and early adoption programs to support. A federal agency would serve as a gatekeeper to certify the acceptability of investments that could be proposed to the RIIBs. RIIB funding would derive from a surcharge on electricity sales. The surcharge funds would be allocated to the RIIBs by a set of state-level trustee organizations. The RIIBs would have to compete to win enough support from enough trustees to fund their portfolios. This regionally based demonstration and early deployment funding mechanism (shown schematically in figure 5.1) fulfills our principles more fully than the other proposals that we have considered.

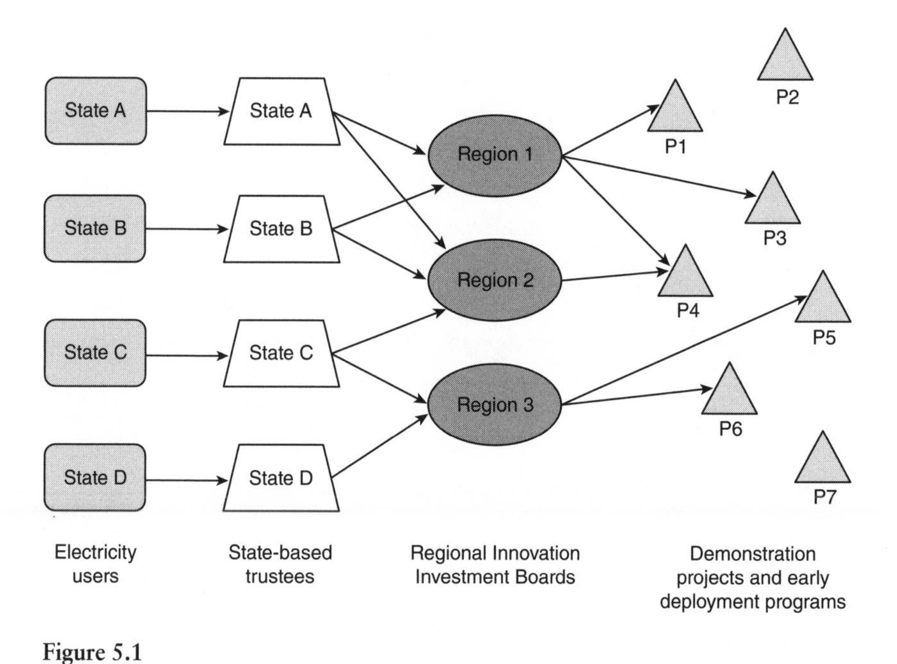


**Figure 5.1**

A regionally based mechanism for funding energy innovation demonstration and early adoption.

## Regional Innovation Investment Boards

**Recommendation: Congress should authorize the creation of a network of regional innovation investment boards that would invest in demonstration projects and early deployment programs for low-carbon energy technologies.**

RIIBs are at the cutting edge of the energy innovation system that we envisage. Each board's membership would comprise the electric power companies in that region, including "smart integrator" utilities, transmission companies, vertically integrated utilities (in those parts of the country where they remain in place), independent power producers, providers of grid services, and demand response and energy service firms. The membership would thus reflect the perspectives of innovation users with close knowledge of the electric power system and the requirements of the electricity market. Some of the members would be potential buyers and operators of the innovations being demonstrated and deployed; others would be indirect users of them. For example, an independent power producer and a smart integrator might both be represented on a board that is supporting a project to demonstrate a new central station generation technology. The IPP would be an eventual buyer of the technology, while the smart integrator might ultimately incorporate the technology into its network. RIIB members would elect the governing board, which would have ultimate authority over innovation investment decisions. To limit conflicts of interest, representatives of entities proposing projects to an RIIB would not be eligible to serve on its governing board.

Projects that might be funded by the RIIBs would include first-of-a-kind large-scale demonstration projects and "next few" post-demonstration projects. A variety of investment instruments could be used, including loans, loan guarantees, interest buy-downs, or equity infusions. Investments by the RIIBs would leverage larger amounts of private-sector funding for these projects. The RIIBs might also invest in programs designed to accelerate the early take-up of innovations in distributed generation. The investment instruments used for this purpose could include large-scale user subsidy programs such as feed-in tariffs, rebates, or subsidized loan programs. (Chapter 6 discusses the role of RIIBs in other parts of the electricity system, such as the smart grid.)

Many different kinds of teams might propose projects and programs for RIIB support. In some cases a single entity, such as a technology vendor, could propose a demonstration project. A state government could propose an early adoption program. Alternatively, the project team might be a consortium of established and entrepreneurial vendors, other members of the technology supply chain, independent power producers, and national or academic laboratories. Those proposing projects would seek RIIB funding not as the sole source of finance but rather to augment their own investments and to lower their risks. RIIBs would choose among competing project proposals based on the strength of the project team, the quality of project management, the attractiveness of the technology, the extent of self-funding by the proposers, and so on.

To seek RIIB financing, projects would first have to be certified by the federal gatekeeper organization (criteria for certification are discussed below). Certified projects could be proposed to one or more RIIBs for funding. RIIBs could operate independently, or they could collaborate with each other on investment decisions. They would be obligated to commit all funds received in a given year by the end of the following year; in other words, they would not be permitted to accumulate funds. However, funding commitments to projects could be spread over time, with future payments made contingent on the satisfactory performance of the project team.

The boundaries of the RIIBs would need to be carefully drawn. To avoid excessively fragmented decision making, no more than ten such boards should be created. One approach would be to build on existing groupings of states with prior experience working together on regional energy and environmental issues—for example, the New England states, the Northwest states, and the Southwest states. A somewhat different approach would be to use the boundaries of the eight North American Electric Reliability Corporation (NERC) regions (see figure 5.2). Each NERC region is relatively autonomous and has a distinctive character that reflects its history and demand profile.[14]

Mapping the RIIBs onto the NERC regions or some other regional structure would have the benefit of steering innovation activities toward the energy transition pathways most suitable and attractive to each region of the country. Over time, each RIIB might be expected to specialize in innovations of particular interest to its region—for example carbon capture and storage in the coal-dependent Midwest, nuclear in the Southeast, solar in the Southwest, and offshore wind in the Northeast. A RIIB might also tend to invest in projects located in its own region, though it would not be limited in this regard.

## State Trustees

**Recommendation: Federal and state legislation should establish an innovation surcharge on retail sales of electricity and create trustee organizations in every state to allocate the proceeds to the RIIBs.**

Funding for the RIIBs would derive from an innovation surcharge on retail sales of electricity. Individual RIIBs would not have an entitlement to funding, however. Rather, disbursing the proceeds from the surcharge would be the responsibility of a trustee organization in each state. After a start-up period in which each trustee would be required to direct all surcharge funds collected in its state to the RIIB in its region, trustees would be free to allocate funds to one or more RIIBs in other regions. The allocation would be based on each trustee's assessment of which

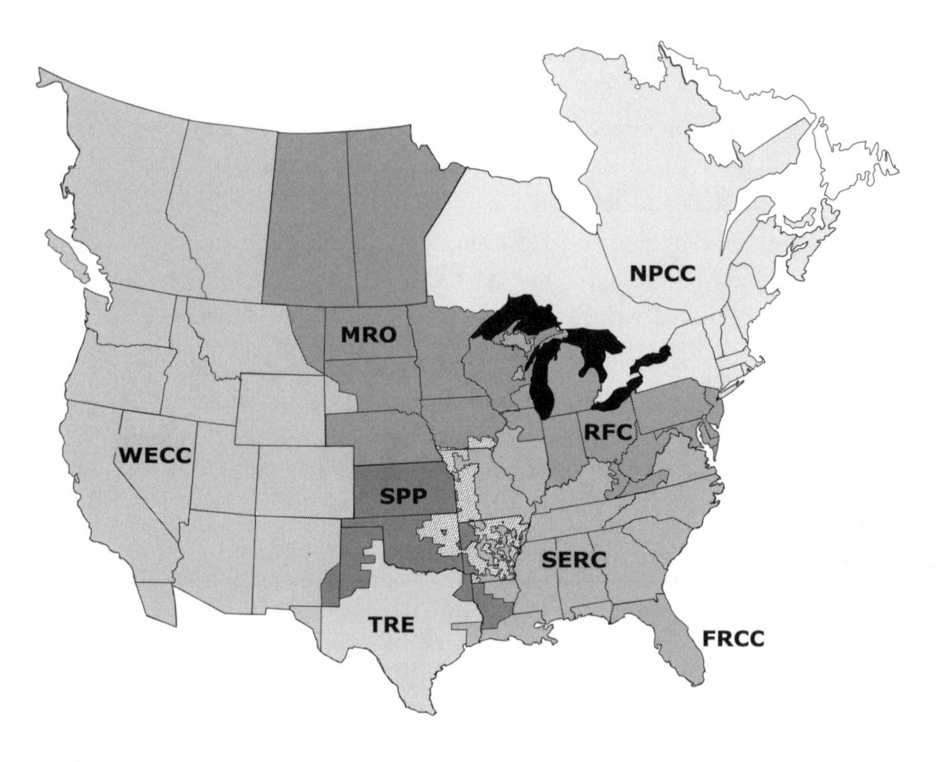


**Figure 5.2**

North American Electric Reliability Corporation (NERC) regions. *Source:* NERC.

RIIB portfolios of demonstration projects and early adoption programs most closely matched its needs. The trustee organizations would be more broadly representative of stakeholders in the electricity system than the RIIBs themselves. Their members might include a variety of business groups, government organizations and officials, environmental and labor groups, and technical experts. Depending on the state, trustee board representatives might be appointed or elected. Trustees would have to allocate all proceeds from the surcharge to RIIBs within a year of receipt.

We suggest an innovation surcharge of 0.2 cents per kilowatt hour. This would add roughly 2 percent to the average retail price of electricity in the United States, or about $1.80 per month to the average residential electricity bill, and would generate roughly $8 billion per year. (As noted, these funds would leverage private sector funds such that the total investment induced might be at least two to three times larger.) This funding stream is similar in scale to that proposed by the President's Council of Advisors on Science and Technology to support federal energy RD&D.[15] Federal legislation would be required to enact the proposed surcharge. It might justifiably be enlarged. Revenues from other sources, such as revenue raised by eliminating tax incentives for oil and gas production or by imposing a carbon tax, could be added as well.

## A Federal Gatekeeper

**Recommendation: A federal gatekeeper should certify proposals for demonstration projects and early adoption programs before they become eligible for RIIB investment.**

The third new institution in the regionally based funding scheme that we propose is a federal gatekeeper organization that would certify, decertify, and recertify projects and programs. Every proposal would have to be certified before it could be presented to the RIIBs. The gatekeeper would make sure that RIIB investments were supporting national purposes. All proposals would have to demonstrate the potential to lead to significant reductions in carbon emissions at a declining unit cost over time. The gatekeeper would apply this criterion to all proposals in determining their eligibility for RIIB funding, regardless of the nature of the innovation under consideration. The gatekeeper would not determine whether a specific proposal should receive funding, nor would it rank innovations or evaluate the organizational capabilities of the proposing organizations. These tasks would be undertaken by the RIIBs themselves.

Projects or programs would need to meet a minimum size threshold (say $500 million of public and private funding combined) to qualify for RIIB funding. The objective for first-of-a-kind projects would be to (1) demonstrate technical feasibility at full scale and (2) generate and disseminate data on the technology's cost, reliability, and environmental performance. "Next few" post-demonstration projects would need to incorporate information from previous demonstrations with the aim of making continued rapid progress toward competitiveness with incumbent technologies. Large-scale user subsidy programs, similarly, would need to show their impact on technological progress down the learning curve.

The gatekeeper would be responsible for monitoring progress. Proposals would be judged in part on how effectively they put pressure on innovators to exploit learning to reduce costs. Public subsidies would have to decline steadily as experience with the innovation is gained. Certification would be granted only for a limited period—five years, say—and could be withdrawn at that time if progress proved too slow. The gatekeeper would also track projects and programs that target the same innovations to guard against duplication and overlap. At the same time, it might take into consideration the value of pursuing several different approaches in parallel as circumstances warrant.

The gatekeeper would need to be a technically competent organization capable of representing the broad public interest. It could be part of an existing federal agency such as DOE or it could be a stand-alone unit. It would need to be capable of assessing the technical progress of projects and programs. It would also need to be able to evaluate the potential of scale economies and future learning opportunities. Finally, it would have to have a global perspective and be knowledgeable about developments overseas, so that RIIB investments would not simply duplicate work being done elsewhere.

## Pros and Cons of the RIIB Approach

A regionally based funding mechanism offers several advantages over current practice. It would create a dedicated new stream of funding for what has been a chronically under-resourced part of the energy innovation system. It would avoid the stop-and-go pattern associated with the federal appropriations process, where decisions are too often driven by short-term energy price volatility, technology fads, and election results, and would instead generate the steady, predictable flow of funds needed to make credible multiyear investment commitments. The innovation surcharge would generate roughly $8 billion per year and would induce as much as $20 billion per year of total investment in demonstration and early-adoption activities—enough to have a major impact on the nation's energy innovation challenge, without increasing the federal budget deficit.

By putting project and program selection in the hands of potential innovation users—the RIIBs—rather than in the hands of government officials or elected representatives, this new institutional framework would be much more responsive to feedback from the market and from technical results than the existing system. At the same time, the public interest would continue to be strongly represented by the federal gatekeeper and state trustee organizations.

The institutional framework we propose would also introduce multiple levels of competition into the innovation process. In the past, demonstration projects have been selected through a highly centralized and sometimes arbitrary process, in which individual congressional champions or national laboratories have often played influential roles. In the new arrangement, project teams, once certified, would compete with each other for funds from one or more RIIBs to design, construct, and operate demonstration and post-demonstration plants, or to implement early adoption programs. The RIIBs, in turn, would compete with one another to secure support for their portfolios from the state trustees. A regional board with a portfolio deemed promising by multiple trustees would see its investment budget swell, while those with less promising portfolios would shrink. This more decentralized scheme would also allow new entrants who lack connections to the existing federal R&D structure to get a better hearing for their ideas than they do at present.

Finally, our scheme would create opportunities for regional needs and preferences to be expressed in the energy innovation system, and would give states a direct stake in innovation outcomes.

Our scheme also has several potential drawbacks. The electric power industry has not been known for its vision and imagination, and assigning investment responsibilities to its representatives might increase the likelihood of cautious, incremental decision-making. The gatekeeper is designed to ward off this possibility and focus the RIIBs on significant reductions in carbon emissions at declining unit cost. At the same time, the injection of conservative user perspectives at the demonstration and early adoption stages has the virtue of constraining politically popular projects and programs that are not consistent with the overall goal of reducing risks to a level that would allow private investors to fund future projects on their own.

A second potential drawback involves resistance from entrenched interests. The DOE laboratories, for instance, have long enjoyed a privileged position in the U.S. energy innovation system. They exert a strong influence over DOE resource allocation decisions and receive by far the largest share of DOE RD&D funds.[16] In our proposed framework the laboratories would have to compete with industry, universities, and others for a share of the RIIB energy innovation funds, although the total amount of funding available would be much greater. (The labs' traditional role in support of other DOE mission areas, such as national security, basic science, and environmental quality, would not be affected.) States might also have difficulty accepting the regional approach. There could be pressure on the trustees to spend funds collected within the state only on projects within the state. However, the proposed scheme would not permit the trustees to fund individual projects. Rather, they would be required to allocate their funds at the portfolio level, with each RIIB portfolio typically comprised of multiple projects distributed across states both inside and outside the RIIB's region. In addition, the RIIBs would be membership organizations and therefore less susceptible to this kind of pressure.

Perhaps the most serious objection to the scheme we propose is that it would entail the creation of many new organizations and take time to set up. There will be a learning curve for each of these organizations. That is unavoidable, but the process should be complete in a period of years, while the payoffs will occur over several decades. Competition among the RIIBs and discipline provided by the gatekeeper should minimize the risk of administrative inefficiency over the medium-term.

An illustration of how this scheme would work can be found in box 5.2 on "The RIIB-Gatekeeper-Trustee Innovation System in Practice: The Case of Nuclear Power."

**Box 5.2**

The RIIB—Gatekeeper—Trustee Innovation System in Practice: The Case of Nuclear Power

To see how the regionally based system for demonstration and early deployment might work in practice, consider the case of nuclear power, which—perhaps more than any other major energy technology—has depended on and been influenced by direct federal involvement in the past. High capital costs for nuclear power plants and the lack of a resolution to the problem of nuclear waste disposal have been the main obstacles to a revitalization of the U.S. nuclear power industry in recent years. More recently, the Fukushima nuclear accident has brought a renewed focus on safety concerns, with uncertain implications for the industry's near-term prospects in the United States and elsewhere. Yet many innovations in nuclear reactor and fuel cycle technologies are under development that have the potential to contribute to safety, environmental, economic. and security objectives.

Historically, the federal government has played a leading role in commercializing nuclear power reactor and fuel cycle technologies. In recent years its primary focus has been to promote, in collaboration with the nuclear industry, the development and deployment of advanced light water reactors. Relevant federal policies have included regulatory reforms, such as the more streamlined combined construction and operating license procedure set in place by the U.S. Nuclear Regulatory Commission two decades ago; federal cost matching to encourage private firms to seek approval for advanced reactors under these new licensing procedures; energy production tax credits covering the first 6000 MWe of new nuclear capacity; federal regulatory insurance to cover the cost of regulatory-induced delays in the operation of the first six new nuclear power plants; and federal loan guarantees to support advanced nuclear power plant construction.

In addition, since the early 1980s the federal government has been legally responsible for the final disposition of spent fuel discharged by current as well as any future nuclear power plants. The government has also provided significant funds for the development of advanced nuclear energy technologies, including sodium-cooled fast reactors and gas-cooled high-temperature reactors, as well as a range of new fuel cycle technologies, especially for spent fuel reprocessing, actinide separation, and transmutation. Although the government has played no role in developing advanced uranium enrichment technologies since the privatization of federal enrichment operations in the 1990s, in recent years it has offered loan guarantees and other funding to first-of-a-kind enrichment plants planned by private firms.

Few of these policies have been successful. The federal loan guarantee program, potentially the most important of the government's promotional measures, has been largely ineffective so far. Seventeen utilities have submitted applications for these federal loan guarantees, on behalf of more than 20 advanced light water reactor projects. But in the five years since the loan guarantee program was first introduced, DOE has approved guarantees for only one new project, and most of the utilities have either withdrawn from the program or suspended their plans.[a] The economic viability of those projects was adversely affected by the decline in natural gas prices and by the unwillingness of the Congress to enact a carbon price. In some cases, however, DOE's loan guarantee office and the Office of Management and Budget were simply unable to agree on acceptable terms with the utilities.

The prospects for nuclear development have also been adversely affected by problems in the government's program for high-level waste disposal. The latest setback was the Obama administration's decision to halt work

**Box 5.2 (continued)**

on the Yucca Mountain nuclear waste repository project, which for the last twenty years has been the sole focus of federal spent-fuel disposal activities. The current impasse over Yucca Mountain, the lack of an alternative disposal plan, and the continued inability of the federal government to fulfill its contractual obligations to remove spent fuel from commercial power reactor sites—a long-running irritant in relations between the nuclear utilities and the government—together convey an impression of government ineptitude and ineffectiveness in a domain vital to the future of the nuclear industry.

Over the past year DOE has begun to promote the accelerated development, licensing, and deployment of small, modular nuclear power plants, both light water reactor (LWR)-based concepts and non-LWR designs. With congressional support, DOE is proposing to transfer some funds from other nuclear energy programs to this new initiative. But it has not yet indicated how it will allocate available federal funds among design certification and licensing activities, demonstration projects, deployment incentives, or the creation of reactor manufacturing capacity. Nor has it said how it will direct these funds among several privately developed small reactor concepts.

The arrangements we advocate in this chapter would transform radically the U.S. approach to nuclear energy innovation. DOE and its laboratories would continue to conduct R&D in nuclear reactor and fuel cycle technologies, including R&D in the important areas of safety and regulatory development. But their role in commercializing new nuclear technologies would be diminished. RIIBs, responding to proposals by nuclear innovators, would be the primary decision-makers on key strategic questions: how to manage the financial risks of new advanced light water reactor power plant construction; what the appropriate division of effort should be between the implementation of incremental advances in large light water reactors versus the commercialization of small modular reactors; which particular reactor technologies and commercial groupings are most deserving of demonstration and "next few" investment; how to allocate investment among different approaches to new reactor and fuel cycle technologies, funding human capital development, and so on.

Ironically, the federal high-level nuclear waste disposal program, which is almost certainly the most egregious example of a nonperforming federal investment in nuclear innovation, has been financed for decades by a revenue-raising mechanism—a 0.1 cent per kilowatt hour fee on nuclear electricity paid into a federal Nuclear Waste Fund—with some similarities to the arrangement we advocate here. Some $25 billion has now accrued in the fund (including interest). But the differences are important. All decisions regarding the use of the Nuclear Waste Fund have been centralized within the DOE and overseen by Congress. Federal spending in this area has been subject to extraordinary constraints, such as the nearly twenty-five-year-long statutory prohibition on investigating disposal technologies and geologies other than those at Yucca Mountain—a provision that actively discouraged innovation in the field of nuclear waste disposal. Congressional oversight, exercised through the annual appropriations process, has routinely been politicized as opposition to the Yucca Mountain project has mounted, and DOE has frequently complained about its inability to implement plans for the site because of congressionally imposed spending restrictions. The nuclear utilities and their ratepayers have had essentially no recourse in the face of these problems, other than to sue the federal government for its failure to meet its contractual obligations to remove their spent fuel. We concur with the recommendation, first made more than thirty years ago, to move responsibility for implementing nuclear waste management and disposal out of DOE and into an independent government authority.[b]

a. For a nuclear industry perspective on the federal loan guarantee program, see "Credit Subsidy Costs for New Nuclear Power Projects Receiving Department of Energy (DOE) Loan Guarantees: An Analysis of DOE's Methodology and Major Assumptions," Nuclear Energy Institute White Paper, August 2010.

b. The latest recommendations along these lines have been made in a recent MIT study, *The Future of Nuclear Fuel Cycle* (Summary Report, Massachusetts Institute of Technology, 2010). The original recommendation was made in Mason Willrich and Richard K. Lester, *Radioactive Waste: Management and Regulation* (New York: Free Press, 1977). A scheme similar to the one proposed here for energy innovation might also be considered for managing high-level nuclear waste.

## Conclusion

The problem with past energy technology demonstration projects was not that they failed but that at some point the goal became to avoid failure. For the leaders of these high-profile projects and their supporters in and outside government the costs of failure were too great, so failure had to be avoided at all costs. But some of the strategies for preventing failure themselves proved costly, including driving out other alternatives prematurely, refusing to recognize legitimate problems until long after they arose, and failing to acknowledge that key assumptions were no longer valid. And, of course, these projects also generated a constellation of opponents, whose goal became precisely to cause their failure, and to prevent them from producing anything useful. In this environment, the most important goals of the innovation process—generating new information and learning quickly about the strengths and weaknesses of alternative approaches—were undermined.

For large-scale technologies, developed in government-led and government-financed projects, this kind of pathology is an ever-present risk. Yet the rapid development and deployment of new large-scale electric power technologies will be essential to the low-carbon energy transition. So a critical task is to devise an innovation system in which, even for these large-scale technologies, multiple pathways can be pursued and failure is tolerable. In this chapter we have suggested one such scheme. Undoubtedly others can also be devised and we encourage further explorations along these lines. The most important goal is to create an institutional structure that can accommodate and promote diversity, experimentation and competition in the innovation process—even for large-scale technologies and even during the downstream stages of demonstration and early adoption.

The second wave of innovation is not only about large-scale electric power technologies, however. As we shall see in the next chapter, there are many other innovations that should be pursued in the second wave. For these, too, new institutional structures are needed to unlock the full creativity and competitiveness of America's innovators.

# 6

# The Second Wave of Innovation—Part 2: The Rest of the Electricity System[1]

Central-station generation supplies almost all of the electric power consumed in the United States today, and it will continue to be a major part of the decarbonized electricity system decades hence. Economies of scale drove the development of big power plants and the big transmission lines that connect them to customers. These technologies were more affordable than smaller-scale, more widely distributed power systems. Although these economies of scale have moderated in recent years, they have not disappeared. Even for modular low-carbon energy supply technologies, such as solar photovoltaics or wind turbines, there are cost advantages to deployment in large arrays.

This conclusion drives our emphasis on innovation in central-station generation, and it is the reason that we devoted much of chapter 5 to the subject. But one cannot simply plug twenty-first-century forms of central-station generation into the remains of the twentieth-century electrical system and be done with it. One obvious issue, touched on in chapter 3, is that locations that are well suited to fossil fuel–fired power plants are not necessarily the right ones for their low-carbon replacements. Coal plants, for instance, benefit from access to waterways, whereas wind farms should be sited (where else?) in windy places. Shifting electricity production to new locations will demand new transmission facilities linking those locations with cities and other load centers.

A more subtle set of issues pertains to the quality of electricity and is well illustrated by many forms of renewable generation. When passing clouds block the sun, or when the wind is not blowing, the output of systems that are powered by these sources drops precipitously. But demand generally does not drop—most lightbulbs need to stay on. In order to keep the system balanced at such moments, another source of power has to step in or some less sensitive source of demand has to shut off. Intermittency of this sort can be predicted to some degree, but not perfectly. As the share of intermittent generation grows, managing the grid becomes increasingly complicated. "Smart integrators" will have to innovate as a result.

So, even if the aim were to shift to low-carbon central-station generation on the cheap by retaining the rest of the old power system infrastructure, this would not be feasible. In fact, it would not even be desirable. If the entire inventory of existing fossil-fueled power plants could be retrofitted at a stroke with carbon capture and sequestration, thereby creating the option of keeping the rest of the system as is, the country would still be better off upgrading it. Innovations outside the realm of central-station generation that offer potentially huge advantages are on the horizon.

Some of these innovations, in fact, may diminish the overwhelming dominance of central-station generation. Distributed generation, grid-scale storage, and the smart grid all promise advantages in terms of reliability and resilience. If the costs of these technologies come down, and if they can be integrated effectively into the power system, their roles will grow significantly. And if that happens, it is very likely that customers who are now passive recipients of electricity service will start to play a much more active role in power networks and even contribute to innovation in them.

As we saw with central-station generation and energy efficiency, however, institutions will need to be restructured to bring these innovations to fruition. Utilities must be prepared to receive distributed power and to manage the fluctuations in supply and demand that would result from innovations on the customer's premises. Power pricing must be modified so that households and establishments have the incentive to adopt innovations that allow them to reduce and better manage electricity demand and to make the use of these innovations habitual. Public policy can play an essential enabling role in this process through regulation, standard-setting, information-dissemination, and support for research, development, and demonstration.

## Opportunities for Technological Innovation

This chapter generalizes about a diverse set of technological opportunities. Low-carbon distributed generation can involve a variety of technologies at a variety of scales. Grid-scale storage may be physical or chemical, thermal or mechanical. The smart grid, both on the customer's premises and at the utility's facilities, will include sensors, communications, decision support, and more. The theme of the chapter is not how such technologies work individually but rather (1) how they might work together with central-station generation to form an integrated low-carbon electricity system and (2) how the U.S. innovation system can help to bring about that outcome.

That said, it is still useful, without attempting to provide a comprehensive catalog of technological options, to give a sense of the components that might be integrated. Distributed generation can power loads ranging in size from a single residence to a small community. Off-grid applications in remote locations like national parks are perhaps most familiar. For some users, independence from the grid is an attractive feature of distributed systems. However, this approach comes with its own vulnerabilities, notably the lack of back-up power in case of system breakdowns. Grid-connected distributed generation systems, by contrast, make it possible to tap the grid for power when the distributed system is down and to "export" power to the grid when on-site electricity production exceeds on-site demand. For intermittent sources like solar photovoltaics and wind turbines, grid-connection alleviates the need for storage, which is usually expensive. Not all distributed generation systems are intermittent; low-carbon power systems that are similar to conventional fossil fuel generators (in the sense that they can be ramped up or down at will) can also be deployed on a small scale. (Examples include some kinds of fuel cells and biomass-fired systems and, in some locations, hydro and geothermal power.)

The best-known form of electricity storage is a battery. Batteries are wonderful for the small applications with which everyone has experience, but where power is needed in larger quantities, they are heavy and expensive. Making them lighter and cheaper is the most important technical objective for developers of the electric car.[2] Storing enough power to keep a small city running is also difficult, although the storage system can be stationary in that case and its weight matters a lot less. Indeed, pumped hydro storage systems, which store energy by pumping water uphill from one reservoir to another and recapture the energy by letting the water run downhill and drive a turbine, have been around for a long time. But sites for pumped hydro are not widely available, and they are not always the right size to meet local storage needs. Scientists and engineers are therefore exploring a number of possibilities for grid-scale energy storage, including the use of new materials for batteries; novel thermal storage concepts, like reservoirs of molten salt; and old ideas, such as flywheels and compressed air. The technical requirements for grid-scale storage vary widely, from helping to match the power output of a group of generators with night-day demand cycles to smoothing out the effects of sudden lulls in wind speed on the output of wind generators.

Another large set of opportunities for technological innovation is embodied in the term "smart grid." "Smart" means that these technologies integrate information and communication technologies with power technologies. They can tell operators with great detail and in real time what is going on in a system, whether the system is a small home area network or a big regional high-voltage network. The smart grid also allows network operators, who might be homeowners or building owners or grid operators or third-party service providers, to control their systems in an equally precise fashion. Of course, such systems are really only as smart as the operators (or software) making the decisions. Who will make such decisions and on what basis are important open questions that we take up later in this chapter.

Although the term "smart grid" suggests a single system, we find it useful to distinguish between the part of the grid that operates "behind the meter," on the customer's premises, and the part that operates "in front of the meter," at the utility's transmission or distribution facilities. A "smart meter" sits at the boundary between the customer and the utility, interconnecting the two networks. The "customer-side smart grid" gathers information about electricity consumption by HVAC systems, lighting, appliances and other devices, storage in electric cars, and generation by rooftop or other systems on the customer's premises. The customer-side smart grid also collects prices, weather conditions, and other relevant inputs from the utility side of the meter. By analyzing all of this information and conveying what it finds in an accessible format, the customer-side smart grid permits more informed decision making about whether to use electricity, when to use it, and how much to use. The smart grid can also implement these decisions by remotely controlling devices in the network.

From the perspective of the smart integrator utility, each smart meter represents a load that must be supplied (or in the case of distributed generation or storage, a source of power for the grid). The level of the load fluctuates continually according to decisions made on the customer side. In addition to communicating with millions of smart meters, the utility-side smart grid tracks power flows and the condition of equipment from sensors located throughout the utility's own distribution system. It also connects upstream to similarly outfitted transmission lines and to wholesale generators and other market participants. Automated controls and decision support allow the utility to manage frequent changes in demand enabled by the customer-side smart grid and intermittent supply from distributed and renewable resources. The utility-side smart grid also allows operators to anticipate outages and to adapt quickly to unanticipated disruptions, enhancing the system's reliability and resilience.

The smart grid as a whole, then, facilitates continuous, efficient adjustment of electricity supply *and* demand in response to changing information. One major focus of smart-grid innovation is "peak shaving." All locations experience daily and annual peaks in electricity demand—that is, hours of the day and days of the year when demand is substantially higher than the rest of the time. During these peak periods, such as summer hot spells, the most expensive power plants are brought on line; when the demand peak passes, they are shut down again. Building and maintaining this "peaking" capacity is very expensive. By arming operators with better information about the kinds of loads that the system might have to meet at the peak, the smart grid enables them to design programs that reward customers for reducing their demand during peak times. Sometimes the utility itself directly controls the reduction with the customer's agreement. This kind of "demand response," along with efficiency improvements more broadly, avoids investments in peaking capacity.

It is important to bear in mind, however, that peak shaving does not necessarily result in reductions in carbon dioxide emissions. If peak shaving shifts the load to another time (rather than eliminating it altogether)—if it simply means, for instance, that a washing machine is run at night rather than during the day—the emissions impact depends on what kinds of generating systems are running at each time. If peak power during the day is provided by a natural gas plant, as is common throughout much of the country today, and nighttime baseload power is provided by a nuclear plant, then emissions go down. But if peak power is provided by a solar system and nighttime power by a coal plant, then emissions go up.

Along with everything else, the smart grid has the capacity to allow decision makers to take into account the consequences of particular choices for greenhouse gas emissions. So we return to the open questions noted above: Who decides? And on what basis? The technological innovations that we have described so far create promising opportunities to accelerate the energy transition. But that promise will not be fully realized unless other kinds of innovation are accelerated as well.

## Open Architecture

The most important institutional innovation that will enable distributed generation, energy storage, and the smart grid to fulfill their potential is the same one that that we have stressed in earlier chapters: the vertical disintegration of the electricity industry. We envision an industrial structure in which many parties compete to provide many kinds of services to the smart integrator and to customers. For such a structure to work, there must be a clearly defined boundary around the smart integrator and standardized interfaces that cross the boundary so that active customers and third-party service providers have the chance to "plug and play" their applications (such as distributed generation, storage, and demand response) with relative ease. This open architecture will accelerate both early adoption and improvements-in-use for the technologies under consideration in this chapter.

**Recommendation: Federal and state policies should foster an open architecture for electricity distribution grids.**

For distributed generation technologies that can supply power and ancillary services to the smart integrator (and in the future, for storage technologies as well), the architectural challenge is to put these supplies on an equal footing with similar resources that flow through the transmission grid. Distribution grids may need to be upgraded to perform functions in two directions that currently occur only one way. Systems that now only pass power downstream to customers will have to be equipped to take power from them as well and to deal with fluctuations in distributed sources of power. "Net metering" is an example. Rather than measuring only power flowing from the grid to the customer, net metering measures power that flows from the customer to the grid as well. Net meters may also record when this power is supplied as well as the value of the power that the grid would otherwise have had to supply at that time, so that distributed generators can be compensated properly. (We return to the important issue of time-varying electricity pricing below.)

In addition, the technical procedures by which distributed generators connect to the grid must be made transparent and easy to follow. As of March 2011, more than forty states had adopted some form of interconnection standard or guideline and net metering requirement for distributed generation.[3] But these rules and guidelines vary widely, may not be mandatory, and do not necessarily apply to all distribution systems; munis and coops, for instance, are usually excluded. Most limit the size of systems and classes of customers that can interconnect. And they are sometimes unduly burdensome or implemented in slow motion by the utility. But palpable progress has been made in the last few years, in both the coverage and quality of interconnection and net metering policies.[4]

The architecture that will govern the customer-side smart grid is less well developed. One way to understand the issues is to think about just how smart the smart meter should be. The dumbest smart meter would record electricity consumption information at short intervals and transmit it to the distribution utility. It would still be much smarter than the traditional meter, from which a usage level is manually collected by a meter reader once a month. The smartest smart meters, however, can act on information that they have collected, instructing electricity-using equipment and appliances to turn on and off when specific criteria are met (when power is cheap or expensive, for instance, or more or less carbon-intensive) or when the owner sends the signal to do so.

One architecture for the customer-side smart grid would rely on the smartest smart meters and place control over them in the hands of the distribution utility (or a third-party service provider working for the utility). The utility or third-party service provider would be able to reduce peak loads and limit power costs by using smart meters, with the customer's prior agreement, to turn off pool pumps, water heaters, and other non-vital equipment and appliances. This approach would provide certainty to the utility that it can manage demand when it is most important (and most valuable) to do so. That certainty, in turn, ought to create a more secure environment for investment decisions about peaking capacity and other infrastructure.

An alternative architecture would use dumber meters and place the smarts in a separate device that is controlled by the customer or by a third-party service provider working for the customer. It is an open question whether, in such an environment, customers would want to be directly involved in decision making using information tools (e.g., web portals, mobile applications and the like) supplied by technology providers, or whether, as many experts believe, they would elect to minimize their active participation and instead make use of automation and third-party service providers. We use the term "customer control" to describe both possibilities, and to differentiate them from the case of utility control. With this kind of architecture, the customer, rather than the utility, would select the criteria that determine when her equipment would turn on and off. If she lives in a location where peak power is provided by low-carbon resources, for instance, she might choose to keep things running when the utility protocol would shut them down. The control device would have to gather the relevant information from the utility, but would have access to many other sources of information as well.

The customer-controlled architecture for the customer-side smart grid maintains a firm boundary around the smart integrator. That boundary would give customers greater freedom to experiment with devices and behaviors than the utility-controlled model and give them confidence that the utility is not acting as "Big Brother" by monitoring and directing their behavior.[5] More important, it would induce third-party service providers to be more innovative in their business models and product offerings by making energy consumers, rather than utilities, *their* customers. Along the way, electricity consumption would quite probably be reduced by more than it would under a utility-controlled architecture.[6]

For example, third-party service providers, including new entrants from the IT industry as well as start-ups and established providers, might offer creative "themes" (like "maximum economy" or "super green") for smart-grid controls that appeal to specific customer segments. They might also offer systems that learn from users and customize themselves in response. Both of these approaches are increasingly important in the

IT industry but are unlikely to be of interest to the utility if it were the customer of the third-party service provider.

The customer-controlled architecture for the smart grid would interface with the Internet through a router, further diversifying the opportunities for innovation. Energy management could be linked through this interface to physical security and maintenance systems for buildings and homes, or controlled through cell phones or the web. The power of social media might be harnessed to allow communities of customers to manage their energy use together. Moreover, the customer-controlled architecture does not preclude the utility from contracting with customers or with third parties who represent customers to provide demand response services, thereby gaining many of the advantages of the utility-controlled architecture.

In sum, we conclude that a customer-controlled architecture for the customer side of the smart grid will lead to more innovation than a utility-controlled architecture. We do not expect that a single decision will cause the country to adopt one architecture rather than the other. There will be many decisions and many decision makers. The federal government has carved out a key role by supporting standards development that will ensure interoperability of components, facilitate data exchange, and address concerns about privacy and security. The National Institute of Standards and Technology (NIST) is leading this process, while FERC is charged with developing rules governing adoption of the resulting standards.

State governments, through their regulation of distribution utilities, will play an even more influential role in determining outcomes. In addition to deciding which markets utilities can enter and on what terms, state regulators oversee utility pricing and investment strategies. Pricing, in particular, is a difficult and contentious issue, and the subject to which we turn next.

## Dynamic Pricing and Other Motivating Information

We have answered our first question—"who decides?"—in favor of customers for many aspects of distributed generation and the customer-side smart grid. Our second question—"on what basis?"—brings us face to face with the problem of setting retail electricity prices. The smart grid allows greater knowledge about and far more precise control of electricity consumption than ever before, but it does not by itself motivate anyone to use these capabilities. A new price structure could do that, and so might other kinds of information about electricity consumption.

**Recommendation: Regulators should approve some form of dynamic pricing, both for customers who have installed smart-grid technologies and for owners of distributed generation who want to sell power to the grid. All levels of government should encourage the provision of other information that motivates customers to adopt low-carbon innovations.**

Most electricity customers pay a single flat rate for each kilowatt-hour of electricity that they consume. That is a necessity in a world of truly dumb meters, since the service provider has no idea when or for what purpose the kilowatt-hours were used. This rate is typically based on the average wholesale price that the provider pays for the electricity (or the average cost of in-house generation and transmission and distribution if the provider is a vertically integrated utility). But there can be a big difference between the average cost and the actual cost. As we have seen, the cost of power at peak hours of consumption—hot summer afternoons, cold winter mornings—is much higher than the average cost. (As noted in a previous section, the reason is that suppliers have to turn on more expensive "peaking" generators to provide enough electricity to meet system demands during these periods.) With a flat rate per kilowatt-hour, however, this is of little concern to the consumer. It is as if he can buy a postage stamp and overnight package delivery for the same price. In fact, unless he is very knowledgeable about electricity, he will never even realize that there were two different products for sale.

In a world of smart meters, it becomes possible to tell the difference between regular postal service and Federal Express. The smart meter knows when the customer is using power, and his control system (whether part of the meter or a separate device) can in principle access the wholesale price at the time of use. If the customer has to pay a retail price based on the actual wholesale price—a typical arrangement for most products in a market economy—he might curtail use during peak periods or shift use to off-peak periods. In the electricity business, this is known as time-differentiated or "dynamic" pricing. According to Ahmad Faruqui of the Brattle Group consulting firm, pilot projects suggest that peak loads might be reduced by anywhere from a few percent up to as much as 50 percent by dynamic pricing. The higher reductions are achieved in pilots that pair dynamic pricing with smart-grid technology.[7] At an MIT seminar in 2009, FERC chairman Jon Wellinghoff went so far as to say that dynamic pricing is the "killer app" of the smart grid.[8]

Dynamic pricing does not necessarily mean that the retail price changes continuously (an approach known as "real-time pricing")—in fact, it probably will not mean that. It might mean that prices change only a couple of times each day, based on the prior history of aggregate demand, as well as on the weekend and seasonally. Or it could mean something in between these extremes. It may be the case, as some have argued, that customers will be paralyzed by the complexity of real-time pricing and would prefer an approach that is simpler and more predictable. On the other hand, some customers might thrive in such a setting, and still others might delegate the decision making to third-party service providers (or electricity retailers) who are well-equipped to handle it. These service providers will compete to discover plans that fit their customers' preferences. The best approach can be worked out through experimentation, competition, and careful evaluation.

Dynamic pricing also has implications for distributed generation and storage. Sellers as well as buyers of power (and of ancillary services) should have an incentive to respond to supply and demand conditions on the grid. Sellers who can supply power at peak periods should be compensated appropriately; in many locations, distributed solar generation (as well as storage) may do so. Distributed generation that competes with baseload power at off-peak times should be less remunerative.

The rates paid to low-carbon distributed generators might reasonably be augmented above market rates in order to drive adoption, accelerate learning, and lower unit costs. Demonstration and early adoption programs of this type would be eligible for support from the regional innovation investment boards described in chapter 5. What we would insist on is that dynamic pricing be incorporated in the design of such programs, so that when public support winds down and subsidies are phased out, market dynamics have already been incorporated into decisions made by investors in the subsidized systems. This is not necessarily the case in state programs that support wind, solar, and other forms of distributed generation today.

Although prices are important, they are not the only basis for decision making. Some customers may want to know about the carbon footprint of power use at any particular time; they might even be willing to pay more for low-carbon power, as some customers already have the option of doing under the "green power" programs currently being offered in a number of utility territories.[9] More powerful motivation may come from historical benchmarks, peer comparisons, and group competitions. Opower, a 2007 start-up, claimed at the time of this writing (July 2011) to have saved more than 400 million kilowatt-hours of electricity by giving customers on-bill comparisons between their energy use and their neighbors', providing energy conservation tips, and other information. The smart grid will greatly enhance the potential for such communication.

A broader public education and labeling program associated with the customer-side smart grid would also be valuable. Third-party service providers and utilities will be a source of much information, but customers do not necessarily trust them. The California utility PG&E, for instance, faced a fierce backlash from customers over the reliability of the smart meters that it rolled out in 2009, as well as complaints about the lack of privacy protection. A federally sanctioned labeling program could help to build public trust in the equipment and offers made by smart-grid service providers.

## Paying for the Customer-Side Smart Grid

Resolving the question of "who decides?" in favor of the customer raises another important question: "who pays?" For distributed generation and storage, the answer is clear and straightforward: the person or entity that owns the asset and receives returns from it should pay for it. In the case of the customer-side smart grid the answer is not as clear, although the same basic rationale applies.

**Recommendation: The bulk of the customer-side smart grid should be paid for by customers, supplemented modestly by taxpayer or ratepayer energy efficiency programs, by RIIB projects, and by more general broadband Internet policies.**

The customer-side smart grid can be divided into three components, which are likely to be paid for differently. The "dumb" smart meter that keeps track of how much power is used and at what time is part of the utility's network, and its costs can be incorporated into the regulatory rate-setting process considered in the next section.[10] The cost of devices to allow remote control of end uses in the smart-grid system will be included in the prices of equipment and appliances. They will add only a small amount to the cost of, for example, an air-conditioning system. What will be quite important is that smart-grid features become standard in all models and that the communications and control protocols are interoperable. These are goals of the NIST effort, mentioned above; in addition, DOE's appliance energy efficiency regulations (discussed in chapter 4) should eventually mandate these features.

That leaves (1) the controller (which might be a software package running on a home computer with a router, or a stand-alone device), a system that has no purpose other than to run the customer's smart grid and therefore represents a modest up-front cost and (2) the recurring cost of Internet access so that the controller can receive prices and other information to support the customer's decision making. There are three potential beneficiaries of this investment. One is the utility. Even though the utility does not control the customer-side smart grid, actions on the customer side may nonetheless allow the system as a whole to avoid investment in new generating capacity. That may reduce the utility's costs, although at least some of these gains will likely be passed on to customers through lower prices. At the same time, the utility must undertake complementary investments on its side of the meter at the distribution level that are costly, as we discuss in more detail below.

A second potential beneficiary is society at large if the customer-side smart grid leads to reductions in $CO_2$ and other power plant emissions. This benefit might justify a taxpayer investment in the customer-side smart grid, like other energy efficiency improvements, all the more so if such an investment drives cost-reducing learning on the part of smart-grid vendors.

The customer herself is the third potential beneficiary. If the smart grid cuts her electricity bills, an investment in the customer-side smart grid could pay for itself. It is true that some customers will be unable or unwilling to shift or reduce their consumption very much in response to the opportunities presented by the smart grid. These customers would not see their bills decline enough as a result of dynamic pricing to justify the investment. Some bills, in fact, will rise. If these customers are using expensive power for a frivolous purpose, such as heating their swimming pools in the dead of winter, the imposition of smart-grid costs on them seems appropriate, because they have a choice and, one presumes, disposable income. If, however, the power is being used for an essential need, like running medical equipment or simply staying warm, the shift to the smart grid may seem unfair.

It is important to be careful in these latter cases to distinguish the problems to be addressed by energy policy from those of other policies. If the problem is the high cost of health care or poverty so severe that the poor must choose between heat and other basic needs, cross-subsidizing high-cost electricity is not the right way to address it. That is what the current system has been doing, inadvertently and out of necessity, through average cost pricing with dumb meters. Low-cost electricity has been made more expensive, and high-cost electricity has been made cheaper. But surely this is a highly inefficient way to tackle problems like health care and poverty. More direct measures to tackle these problems make more sense, whether building on existing government and utility programs that help low-income customers pay their energy bills or through more general poverty alleviation policies.

One review of pilot projects suggests that dynamic pricing will lower the bills of the vast majority (75%–90%) of customers, including low-income customers. Creative rate design and good communication can expand the average savings and the proportion of customers whose bills go down.[11] These findings indicate that the customer-side smart grid will generally pay for itself over time. Refinements in business models, organizational arrangements, and technologies—which we expect as the customer-side smart grid moves through the early adoption and improvements-in-use phases of the innovation process—should solidify this assessment.

Full deployment of the customer-side smart grid will take many years. End-use equipment and appliances are generally long-lived and may not be easily retrofitted for smart-grid integration. The controller provides little value until enough end uses have been upgraded to communicate with it (although the customer might also make manual adjustments to some of her energy using appliances in response to information provided by the control device). Assuming that issues of standardization and interoperability for smart-grid components are addressed and ultimately mandated, nearly universal adoption should be achievable over a period of several decades.

One way to accelerate early adoption and improvement-in-use of customer-side smart-grid technologies might be through business models in which third-party service providers own, lease, finance, or subsidize these systems. Installing technology is a prerequisite for providing service, but it is not necessarily the major part of the provider's cost. It may be worthwhile to use the technology as a loss leader in order to build the service relationship. This business model is the standard one for cell phones, where handsets are routinely given away or deeply discounted as part of service contracts.

Taxpayer- or ratepayer-financed energy efficiency programs might also serve to accelerate customer-side smart-grid innovation. If, as we argued in chapter 4, energy efficiency building retrofits require financial incentives owing to high transaction costs and other unique barriers to later-stage innovation, the same argument may apply to some degree to the customer-side smart grid. For instance, low-income or rural customers may be particularly hard for the customer-side smart-grid market to reach. However, the policy design in such cases ought to require that the most cost-effective energy-efficiency investments get the highest priority. Public investments in weather-stripping and insulation, for instance, would in most cases come ahead of the smart grid in these programs.

The learning investment rationale that we discussed in chapter 2 may also apply to some aspects of the customer-side smart grid. Early adoption projects and programs that seek to accelerate the third stage of the smart-grid innovation process should be eligible to compete for RIIB investments. The RIIBs would decide which of these smart-grid investments would be worth making.

Broadband Internet access is a recurring cost of the customer-side smart grid. If a household is already paying for broadband Internet access, this cost is nil, since the data demands of the smart grid will be small compared to other services, such as video. Two-thirds of the U.S. adult population fell into this group in 2010, according to the Pew Internet and American Life Project.[12] The Obama administration promulgated a national broadband plan in 2010 with the ultimate goal of universal adoption. The smart grid is one of the major applications that justifies this effort.[13]

## Smart Incentives for the Utility-Side Smart Grid

The customer-side smart grid will not work unless there is a utility-side smart grid to go with it. Utilities will need to relay price and system information to millions of dispersed customers, distributed generators, and storage owners. Even more challenging for utilities will be the need to respond effectively when these counter-parties (along with wholesale generators) take action on that information. Smart-grid technology also brings with it opportunities for electricity distribution (and transmission) companies to do better what they already do, such as maintain equipment, ensure power quality, avert problems, protect against physical and cyber attacks, and manage outages.

**Recommendation: Regulators should allow distribution (and transmission) utilities to recover the cost of appropriately justified investments in the utility-side smart grid.**

Outfitting the grid with an extensive array of sensors and developing the ability to process and make use of massive data flows will not be cheap. A recent analysis by EPRI estimated that the net investment needed to upgrade transmission and distribution systems as well as customer premises so as to achieve smart-grid capabilities will be in the range from $338 billion to $476 billion over a 20-year period, or $17 billion to $24 billion per year.[14] This estimate includes the cost of the infrastructure needed to integrate distributed energy resources and energy storage technologies, the digital hardware and software needed to improve transmission and distribution system reliability and security, and the utility-side and customer-side systems needed for full customer connectivity and integration of demand-side resources.[15] The estimate does not include investments in new generation or transmission that may be needed to meet increased demand or accommodate new renewable resources.

These investments for the most part will be made—or not made—by regulated entities. Regulators will ultimately determine which investments are incorporated into the rate base and what rates smart integrators are to be paid for the services they provide. Utilities will only embrace the smart grid if they can realize an economic benefit from doing so. The success of the smart integrator business model will depend in part on shaping utility incentives so as to achieve this outcome.

If and when they do embrace it, utilities will have to make the case for the utility-side smart grid with care and with adequate public deliberation. Recent experience, albeit limited, has not been encouraging. Regulators have at times appeared to be concerned only with keeping prices low in the near term, to the detriment of long-term system performance. Utilities have at times presented unpersuasive or even downright implausible business cases. In one recent instance, a public utility commission turned down a smart-grid proposal even though a significant portion of the cost would have been underwritten by a large federal grant. The commission judged that the ratepayers would still be bearing an unfair share of the costs and risks of the plan, including the risk that the technical standard the utility had selected for the advanced meters it proposed to install would become obsolete.[16]

To be fair, the case for adjusting regulated rates to pay for the utility-side smart grid may be difficult to make, especially in the terms that regulators are accustomed to considering. A customer-controlled architecture means that many of the benefits of the utility's smart-grid investment will depend on complementary investments as well as conforming behavior on the part of customers and their third-party service providers. The "chicken-and-egg" problem of coordinating the two sides of the smart grid may give regulators pause.[17]

Many of the investments that smart integrators will need to make in order to perform key smart-grid functions well are intangible. For example, smart integrators will have to hire well-compensated specialists in fields that have not historically been important in their human resource base. They will have to create management routines and decision-making processes that harness this expertise. Regulators will have to have the sophistication to assess the organizational and institutional dimensions of innovation within the utility sector on top of the technical complexities of hardware and software innovation. They may also have to give some weight to the emissions reduction benefits of these investments (and their ability to help meet renewable energy requirements in states that have them), along with reliability and affordability impacts, as part of their rate-setting processes to support the utility-side smart grid.

We noted these points in chapter 3 and welcomed there initiatives aimed at educating electricity regulators about these issues and cooperation and collaboration among them across state and state/federal lines.

Engaging the broader public in a deliberative process with regard to rate-setting for the utility-side smart grid could also make a big difference. Such a process would help to guard against the perceived or actual "capture" of regulators by utilities, and it has the potential to firm up support for smart-grid innovation more generally. There is a risk that skepticism about particular proposals, much of it well justified, could poison the larger smart-grid innovation process.

## Public Support for Demonstration, Development, and Research

Strengthening the technical case for utility-side smart-grid investments should also aid in building confidence among regulators and the general public. Utilities will have to document the impact of these investments on carbon dioxide emissions, reliability, and affordability in a transparent and open fashion in order to persuade regulators and the public that they are worthwhile. There is a legitimate case for the public to share some of the costs of these investments in the demonstration and early adoption stages.

**Recommendation: Utility-side smart-grid demonstration projects and early adoption programs, and an associated basic and applied research program, should be eligible for cost-shared support through the RIIBs[18] or through taxpayer-funded mechanisms.**

Some aspects of the utility-side smart grid are evolutionary extensions of existing industry trends. These technologies should be relatively easy for utilities to justify to regulators. So-called SCADA (Supervisory Control and Data Acquisition) systems installed on transmission and distribution networks are an example. Relatively dumb smart meters that take hourly readings, along with communications tools to receive their signals, are another. Such meters reduce the costs of meter reading, improve customer service, and allow the utility to forecast its load more accurately. If they allow the utility to achieve a small fraction of the potential demand response benefits, these investments will easily be recouped.[19]

The learning costs of another group of innovations will very likely be covered by private investors, including energy-related information technology applications that are well-suited to venture capital funding. Much of the smart grid (on both the customer and utility sides) will require much less up-front capital and have much quicker paybacks than the innovative central-station generation technologies discussed in this context in chapter 5. They are in technological areas with which the venture capital community is familiar. And they have begun to attract "clean tech" venture capital investments in recent years.[20]

Other components of the utility-side smart grid will be perceived by both utilities and venture capitalists to be too risky to warrant investment on their own balance sheets. In the absence of regulatory approval to recover the costs of these investments, demonstration projects and early adoption programs where costs are shared between the public and private sector may help to surmount the learning cost hurdle. Such projects and programs should be carried out in real-world settings and managed by industry participants who will themselves use the knowledge about costs, risks, and reliability that the projects generate. In the demonstration stage, this knowledge ought not to be proprietary—rather it should be diffused widely to other members of the industry as well as to regulators and the public. In the early adoption stage, proprietary learning should be a central objective. The RIIBs might reasonably consider such investments, or they might fall within the ambit of traditional federal funding programs.[21]

Transparency and openness in the conduct of demonstration projects and early adoption programs may not go far enough in some cases to persuade regulators and ratepayers that their results can be applied to the particular cases before them. These parties, who will ultimately cover the costs of smart-grid innovations, may seek a disinterested source of information about them. This is a role to which the gatekeeper described in chapter 5 is well suited. This body would need to develop a general methodology for evaluating the costs and benefits of smart-grid investments, through an open and inclusive process. It could then apply this methodology to specific innovations that have been demonstrated and certify their effectiveness within specified parameters.

Demonstrating and implementing smart-grid technologies will surely reveal new technical challenges. In many cases, a more fundamental understanding of the electrical system and its components will be important in finding solutions to these challenges. An active research community that is well connected with practice in the field should be nurtured to seek such an understanding. However, power engineering research and education has atrophied in the United States in recent decades.[22] The University of Michigan, for instance, abandoned the field from the 1970s until 2008.[23] EPRI also reduced its program dramatically during the latter part of this period as its members cut back on their research commitments.

Steady federal funding, tied to opportunities to interact with practitioners, will be required to sustain the emerging academic interest in power engineering and to focus this interest in ways that help to serve national needs. In addition to solving research problems, a rebuilt power engineering research community will be a vital asset in training the designers, operators, and managers of the smart grid. Utilities and other industry players will need an infusion of well-trained talent in order to transform themselves into "smart integrators." Linking research and training is a core principle of U.S. science and engineering education, and it should be adhered to in this field.

## Conclusion

Innovation in central-station generation must be complemented by innovation in the rest of the power system for the low-carbon energy transition to be successful. Distributed generation, grid-scale storage, and the smart grid have the potential to make the system less wasteful, more reliable, and more responsive to customer choice. An activated customer base, working with third-party service providers, will help to refine many technologies among the diverse set of possibilities covered in this chapter.

In order to unlock this area of energy innovation, public policy-makers at the federal and state levels will have to take bold steps. Perhaps the most important of these will be the approval of some form of dynamic pricing—the "killer app" of the smart grid—that creates real choices for customers with real value for the power system. Also important will be federal and state actions that lead to an open architecture for the smart grid in which customers, rather than utilities, call most of the shots. Public investments that cover part of the learning investment for innovations in distributed generation, storage, and the smart grid on both sides of the meter will also foster innovation.

These kinds of steps are consonant with the broader principles that we have articulated in this book as well as with the specific recommendations found in previous chapters. A smart grid of the sort that we envision will be tailored to locations and regions, animated by competition in several dimensions, and integrated with the broader energy system. The institutions that surround it will bring new players with new ideas and new demands into the energy innovation process. These new players have the potential to spark a self-reinforcing process that can drive the second wave of innovation.

# 7

# The Third Wave of Innovation: Creating Breakthrough Options

Achieving an 80 percent reduction in carbon dioxide emissions in the next 40 years—what we have called "going from zero to eighty in forty"—will be a remarkable achievement for the United States. It will mean that the nation has recognized and responded in advance to a threat of extraordinary proportions. It will mean that inertia and vested interests have been overcome to create new institutions and build new industries. It will mean that the United States has unlocked the great innovative capacity of its people, firms, universities, and government agencies to bring new technologies, new business models, and new ways of doing things to scale.

The first and second waves of innovation that have occupied the previous chapters can bring America to this point by 2050 if the country sets its mind to the task. The results of these waves will make today's low-carbon energy systems seem as primitive as a 1954 Chevy must seem to a modern automobile designer. But history will not end in 2050. The climate challenge will likely persist and require further cuts in greenhouse gas emissions in the second half of the twenty-first century. If all goes well, the low-carbon energy transition will help the United States deal with other pressing energy issues, such as pollution from coal-fired power plants, the destructive effects of mining and drilling, and the economic and energy security liabilities of heavy dependence on oil imports. But it may provoke unexpected challenges as well that would require a change in the trajectory established during the first half of the century. Society may also place new demands on the energy system to which "business-as-usual" in 2050, whatever that turns out to be, is not well suited.

The U.S. innovation system should therefore be setting a third wave of potential breakthrough options in motion even as the first and second waves of innovation are breaking over the energy sector. This third wave might draw on ideas that seem impractical now, or on science and engineering fields that have not previously been seen as relevant to energy. Alternatively, concepts that have been on the drawing board for a long time but that have never been proven in practice might be realized in the third wave. The innovation system should provide the resources, the physical space, and the intellectual space for creative individuals, teams, and organizations to explore a wide range of technological possibilities, as well as a set of mechanisms to narrow them down, so that a portfolio of new energy innovation opportunities is available to be scaled up between 2050 and 2100.

## Breakthrough Ideas

The exact composition of the third wave is unknowable at this point. The process of creating new options is full of risk and surprise. This year's hot new scientific discovery is next year's trivia answer. Unknown graduate students cook up schemes that succeed beyond their wildest dreams. Still, the following small sample of possibilities provides a hint of what the third wave might bring.

### Carbon-Neutral Biofuels

Today's biofuels industry is mainly based on converting sugar cane, corn kernels or cellulosic materials into ethanol. Technological advances will improve the downstream efficiency of these processes, but the sunlight-to-biomass conversion efficiency of natural photosynthesis in typical crop plants rarely exceeds 1–2 percent, so the land requirements for production at scale are inherently large. (Sugar cane is an exception, with peak conversion efficiencies as high as 8 percent.) Because energy inputs for cultivation, processing, harvesting, and delivery are relatively high, even second-generation biofuels reduce carbon emissions only modestly relative to gasoline.

Third- and fourth-generation biofuels may deliver more dramatic improvements. These new "electrofuels" technologies replace natural photosynthesis with the rapidly developing tools of synthetic biology. They use genetically engineered microorganisms and electrical energy to convert carbon dioxide and hydrogen directly into molecules that are fully substitutable for conventional fuels. For example, Ginkgo Bioworks, an MIT spinoff, is using genetically reengineered *E. coli* bacteria and electricity to convert carbon dioxide into gasoline. DOE's Advanced Research Projects Agency–Energy (ARPA–E) has recently begun funding Ginkgo and a dozen other electrofuels research groups, most of them university-based.[1] Joule Unlimited, another startup, claims to have found a way to harness photosynthesis without having to grow biomass at all. Its process uses a genetically engineered photosynthetic microorganism to convert sunlight, carbon dioxide, and water directly into diesel fuel.[2] Yet another startup, Sun Catalytix, has developed a process that mimics photosynthesis by using electricity and a cobalt-based catalyst to split water into hydrogen and oxygen molecules. These can later be combined in a fuel cell to produce electricity again. If the original electricity is generated by solar photovoltaics, the process in effect stores the solar energy for later use. Alternatively, the hydrogen can be converted into a liquid fuel.

If even one of these biofuel innovations succeeds, it could indeed be a game changer. Success, however, will probably not be known for at least 10 to 20 years, since that is how long it will take to bring the technologies to commercial scale and determine their economic performance. Gaining a significant share of the vast U.S. transportation fuels market would take much longer. Even today, more than 30 years after the federal government first began promoting it, corn ethanol only accounts for about 4 percent of total U.S. vehicle fuel use, and much of that is due to a regulatory mandate.[3]

### Advanced Solar

The cost of solar photovoltaic systems has been falling as sunlight-to-electricity conversion efficiencies inch up and as suppliers pursue economies of scale and process innovations in big new manufacturing plants. But in most locations, electricity from PV is still not close to being competitive with conventional generating technologies without generous subsidies. Progress down the cost curve, driven until now mainly by advances in the active photovoltaic material itself, will become more challenging as cost-cutting efforts begin to focus on the supporting cast of lower-tech system components, which are becoming a larger share of the total cost. The cost of this "balance of system"—that is, the electronic packaging, module framing and support structures, and power conditioning components—along with the cost of installation, are now responsible for two-thirds or more of the total cost for the current generation of PV technologies. Reducing these costs will not be easy, and for the current generation of technologies it will be hard to achieve cost parity with grid-supplied electricity. This would be true even if the light-absorbing photovoltaic semiconductor material turned out not to cost anything at all.

Researchers are now exploring many promising ideas that could overcome this barrier in the longer term. For example, MIT's Vladimir Bulovic has shown how to print a nanostructured layer of light-absorbing quantum dots onto a transparent organic semiconductor. Coating a pane of standard window glass with this material would enable the window itself to provide power for lighting and other devices, and by piggybacking on the installation of the windows, the incremental cost of PV installation would be essentially zero. The main challenge is to work out how to increase the efficiency of the photovoltaic film without diminishing its transparency to visible light, and also to ensure that degradation does not unduly shorten the life of the coating. Another interesting development, also recently demonstrated in an MIT laboratory, is to use printing techniques to coat paper with organic photovoltaic material. This approach could allow the rapid installation of ultra-lightweight solar cells over large areas with just a staple gun.

**Fusion**

For decades thermonuclear fusion has been seen by its supporters as the eventual successor to nuclear fission. Fusion has two main advantages: it does not produce long-lived radioactive waste products and its fuel consists of deuterium and lithium, which are available in essentially unlimited quantities. Impressive progress is being made along two main technological pathways to understand how to tame the extraordinary energy that fuels the hydrogen bomb and the sun itself. In magnetic confinement fusion, hot fusion plasma—the interior of which reaches a temperature of 100 million degrees Celsius—is contained within a doughnut-shaped region by powerful magnetic fields. In laser-driven inertial confinement fusion, intense laser beams bombard tiny frozen pellets of fusion fuel, causing very small thermonuclear explosions several times per second.

Both approaches remain far away from commercial application. The experimental facilities needed to advance the technology are so large and expensive that it will take decades for each learning cycle to unfold—the opposite of the rapid-fire, parallel learning that ideally occurs in commercial innovation. The flagship of the magnetic fusion program, the $23 billion ITER international tokamak experiment now under construction in southern France, is currently scheduled to start operating at the end of the decade and then to run for 20 years. It will generate much useful information, though not enough on its own to allow a commercial demonstration reactor to be built as the next step. (For example, it will not answer many crucial questions about materials performance.) Yet even with seven governments paying the bills, ITER's huge cost, which has more than tripled since planning for the project began, means that there will be little public money available for other important fusion R&D in the meantime.

Development work is now underway on several alternative fusion concepts on a smaller scale, with private investors footing part of the bill. The chances are slim that any one of these approaches will succeed. But if there are enough of them, and if the cycle time for development can be compressed relative to the giant mainstream programs, the overall probability that fusion will actually make a contribution to world energy needs in the second half of the century may be enhanced. Government support for these programs would be a good idea. A government-funded basic technology program involving, for example, facilities to test advanced materials in the extreme environments of fusion reactors could help both the mainstream technologies and the heterodox alternatives.

### "Senseable" Cities

The automobile changed the face of America in ways its inventors never imagined—spawning highways, supermarkets, suburbs, exurbs, and much more. Electricity's impact on human geography has surely been even greater. In both cases technological innovation spurred social innovation, which in turn drove yet more technological innovation. The effect of these interlocking sociotechnical systems on patterns of energy use has been gradual, unforeseen, and immense.

Today the personal and work lives of Americans are being transformed by digital technologies. The larger consequences of these changes are impossible to fully discern. How will our cities evolve in response to them? Predictions are difficult, but Carlo Ratti, director of MIT's SENSEable City laboratory, invites city dwellers to consider a future (not so far off) in which our cities are blanketed with networks of vast numbers of low-cost sensors, collecting information about different aspects of the urban environment, processing it, and acting on it—all in real time. The urban environment, in this scenario, will be talking back to its denizens, perhaps revealing wholly new ways to manage urban resources.

In a neighboring MIT laboratory, researchers are developing a "sensing skin" for the built environment. This thin electronically active fabric could be glued to concrete structures, especially in areas such as the underside of bridges that are prone to cracking, to allow continuous real-time monitoring and early warning of dangerous situations. Elsewhere at MIT, scientists and engineers are exploring ways to engineer concrete at the atomic scale so as to preserve its strength properties while reducing the amount of carbon dioxide that is emitted when it is manufactured—an attractive possibility since concrete production is today responsible for several percent of global carbon dioxide emissions. Across the campus, a consortium is investigating how new computational design, sensing, and fabrication tools could be used to create mass customized flexible living spaces, enabling the delivery of new kinds of services in the home, including health care services, at low cost and with a near-zero energy footprint.[4] The viability of all these developments will hinge on parallel innovations in the organization of the construction and housing industries, whose extreme fragmentation, suspicion of new technology, inefficiency, and penchant for "one-off" projects are major barriers to progress today.

## Expanded, Sustained, and Stable Federal Funding

Breakthroughs like those discussed above are unlikely without sustained research on fundamental scientific and also social scientific problems. Although the beneficial impacts of such advances may not be observable on a large scale for decades, the fundamental research on which they will be based must be funded adequately today. Additional funding for translational research to explore practical applications and proof-of-concept testing is also critical. These are high-risk, long-term bets of the kind that the private sector will not make on its own. In some cases only the federal government can make them on the scale required.

**Recommendation: The federal government should provide stable, sustained funding at a much higher level than in the past for the creation of new, low-carbon energy options for the long-term.**

Our four-stage model of innovation, and particularly the option creation stage, is not well described by current accounting schemes for science and technology investments, so it is not easy to track investment trends in third-wave innovation activities. The definitions used by the National Science Foundation (NSF) revolve around the researcher's intent. If he or she seeks knowledge "without specific applications in mind," the research is classified as "basic." "Applied research" involves "a specific, recognized need," while in development, the effort is "directed toward the production of useful materials, devices, systems, or methods."[5] Similarly, the International Energy Agency's definition of demonstration requires the intent "to help prove emerging technologies that are not yet ready to operate on a commercial basis."[6] Scientists and engineers with all of these intentions may well contribute to each stage of the innovation process. "Improvements in use," stage 4 of the process, may draw on the results of basic research, for instance, while option creation could involve "the production of useful methods."

Nonetheless, even if the fit is imperfect, data that use these definitions are informative with regard to the third wave. They suggest that the past U.S. commitment to creating long-term energy technology options, both private and public, has been small, unstable, and unbalanced. Looking first at the private side, reliable data on energy RD&D spending are surprisingly scarce. The U.S. Energy Information Administration's annual survey of RD&D spending by major U.S. energy-producing companies—most of them oil and gas companies—shows that the spending rate for these companies peaked in 1981, shrank to about a quarter of that level by the year 2000 (after correcting for inflation), and more recently has risen to about a third of the original peak.[7] (These figures use the broadest definition of the relevant investments; much of the spending is almost certainly directed at development, but no breakdown is provided.) On the public side, RD&D spending by DOE peaked slightly earlier, in 1978 (the year after the department was created), and even with the infusion of stimulus spending in the last couple of years, it has not yet recovered to the original level. For particular technology focus areas, like nuclear fission R&D, the booms and busts have been even more severe. The pattern in the much narrower category of basic research, where public spending predominates, has also been fairly volatile, although the long-term trend in DOE's basic research budget has been upward.[8]

Many knowledgeable observers have called for a large expansion of federal funding for energy RD&D. For example, a group of 34 American Nobel laureates called in a 2009 letter to the president for a tripling of total federal energy RD&D spending, to about $15 billion per year.[9] In June 2010, the American Energy Innovation Council—the group of eight U.S. business leaders, including Microsoft's Bill Gates and General Electric's Jeffrey Immelt, to which we referred in previous chapters—made the case that such spending should rise to $16 billion per year.[10] This figure was later endorsed by the President's Council of Advisors for Science and Technology, a nongovernmental group of experts from the academic and private sectors.[11]

There is no definitive answer to the question of how much public investment in RD&D, or in long-term option creation, is enough. But by any measure the level of public funding for energy RD&D has been neither large enough nor consistent enough in the past. We agree that the overall RD&D budget should see a steady rise. But we stress here the importance of making specific budgetary and programmatic commitments to investigate long-term technological options within an expanded RD&D portfolio. As the pressure for near-term results from the energy innovation system builds in the coming decades, there is a risk that research aimed at developing new options with longer (i.e., post-2050) time horizons will be crowded out in public as well as private sector organizations.

## Pluralistic Funding and Public "Spaces"

The availability of a steady stream of federal funding is obviously crucial to sustain the interest of the technical community in long-term research on the low-carbon energy transition. But the conditions under which the work is carried out are equally important. Scientists, engineers, and inventors engaged in the creation of new technological options should have many potential funders that they can approach for support, along with easy access to "public spaces" that facilitate the exchange of ideas.[12]

**Recommendation: Federal support for long-term option creation should be channeled through multiple agencies, allocated via multiple mechanisms, and flow into organizations that foster open information exchange.**

The argument for pluralistic funding rests on the great uncertainty that exists in the option creation stage of the innovation process. Reasonable people, including reasonable experts, will disagree about which projects or technological trajectories are the most promising. Funders have frequently passed up ideas that later proved to be breakthroughs. Only the persistence of the champions of these ideas, and the chance they had to knock on more doors after being turned down, allowed these opportunities to be realized.

DOE will remain a major player in the search for energy technology breakthroughs and, as noted above, its activities in this area warrant additional support. But the Department of Defense (DOD), National Science Foundation, and other federal agencies should also continue to pursue the goal of energy breakthroughs. DOD has a long track record of looking over the horizon to foster new technological paradigms. Its recent commitments to energy innovation, made in response to combat experience in Afghanistan and Iraq, should extend to the early as well as the late stages of the innovation process.[13] The National Science Foundation, too, has recently begun to take a stronger interest in basic research to support energy innovation.[14] While it is important that federal agencies do not duplicate one another's efforts (the problem of duplication can be addressed by coordination from the National Science and Technology Council), parallel breakthrough research programs that reflect each agency's distinctive interests would serve the nation well.

Pluralism in funding styles and methods complements pluralism in funding sources. The National Institutes of Health (NIH) has made enormous contributions to medical treatment through peer-reviewed research. The Defense Advanced Research Project Agency (DARPA) developed many new technologies, including the basic components of the Internet, by relying largely on the judgments of individual program managers. The U.S. Department of Agriculture (USDA) contributed significantly to the rapid growth of agricultural productivity by supporting state-level institutions. DOE, DOD, NSF, and other federal agencies engaged in the creation of new technological options for low-carbon energy should be encouraged to pursue a diversity of approaches to selecting and supporting projects.

From this perspective, the recent proliferation of research programs within DOE is a welcome development, if properly managed, rather than a cause for concern. For instance, DOE's new Energy Frontier Research Centers (EFRCs) represent something of a break from the Department's traditional science funding practices. These centers carry out multidisciplinary scientific research that is intended to address energy technology challenges. The challenges were identified through a deliberative process that engaged a broad spectrum of the technical community. The centers were chosen through peer review and received funding for five years. The EFRC approach complements traditional individual-investigator funding and large-scale facility funding.

On the other hand, DOE's large national laboratories, some of which were founded during the Manhattan Project, retain too strong a hold on DOE's budget and agenda. These laboratories support DOE's missions in national security, science, and environmental quality, as well as energy. They have large, high-quality technical workforces, play an important role in national security R&D, and maintain unique capabilities to build and operate large-scale scientific facilities. The national labs also have strengths in specific areas of energy technology and can assemble large multidisciplinary technical teams to address complex scientific and engineering problems. On the other hand, they play a relatively small role in training the next generation of researchers, which must be a key aspect of the option creation process. They are less able than industry to develop technologies with commercial applications. Moreover, as a 2009 report from the Brookings Institution noted, "the nation's complex energy challenges extend beyond science and engineering into the social and behavioral sciences, professional programs in business administration, law, medicine, and public and environmental policy—all areas where national laboratory expertise is limited."[15]

The National Ignition Facility, the $3.5 billion laser-driven inertial confinement fusion experiment at Lawrence Livermore National Laboratory in northern California, exemplifies the strengths and weaknesses of the national laboratory system. With deep technical and financial roots in the U.S. nuclear weapons program, this scientifically impressive and remarkably ambitious experiment could only have been undertaken at a national laboratory. The main question about laser fusion is not whether it will work technically—although this is a very big question—but whether such an exotic and expensive experiment can ever evolve into a practical machine for the relatively mundane business of generating competitively priced electricity. The project has been justified to Congress partly as a potential breakthrough energy technology but also by its contributions to DOE's other missions of advancing basic science and stewardship of the nation's nuclear weapons arsenal. The voices of potential users of the technology have barely been heard, and the project has been sheltered from a systematic comparison with other early-stage energy options. If, as seems likely, the future of the big and important Livermore laboratory becomes increasingly dependent on the future of the laser fusion program, it may be a long time before such an assessment takes place.

The physical isolation of some of the largest national laboratories has been reinforced in many cases by the information controls and secrecy regulations that are intrinsic to their place at the heart of the U.S. nuclear weapons complex. These aspects of the national lab culture have impeded their ability to serve as open intellectual spaces for debate, argument, and deliberation among a wider community. Universities and even some large corporate labs are more effective in playing this essential role. It is a role that is even more important when viewed in the international context.

## International Collaboration

Averting the worst consequences of climate change is a global challenge. While we have stressed the domestic energy innovation agenda in this book, we are mindful of the consequences of U.S. action (or inaction) for the rest of the world. By maintaining public spaces for open technical exchange, the United States can make an important contribution to the global energy innovation system and spark a two-way flow of potential breakthrough ideas. Stronger international collaboration would intensify this flow.

**Recommendation: The U.S. should strongly encourage researchers working on breakthrough ideas to collaborate with international partners.**

Openness and international cooperation are deeply held values in the research community, especially among scientists. The end of the Cold War and the liberalization of China, India, and other developing countries, along with the global Internet, have allowed these values to be translated into practice more fully in recent years. The National Science Foundation reports, for example, that international coauthorship of scientific articles has risen steadily over the past decade. In 2008, for example, 30 percent of U.S. articles were internationally coauthored.[16]

The same trend is apparent in energy science and technology. The share of papers with international coauthors has risen from 1.4 percent to 14 percent in the past twenty years.[17] In some areas of energy research, nearly all progress is based on international collaboration, owing to the sheer scale of investment required to carry out research projects. The ITER international fusion energy research project is an example.

Still, coauthorship in energy research lags well behind other fields such as nanotechnology and nanoscience. One reason for this lag may be that energy research is held more closely because it is thought to have near-term commercial applications. Another may be that national governments seeking economic advantages discourage international collaboration in this field. As energy becomes an increasingly important priority for national RD&D funding agencies around the world, the danger that "techno-nationalism" will take hold will rise in parallel.

The concern on the part of national governments, including the U.S. government, that the benefits of research that they have funded will be reaped by other countries may be addressed in a number of ways. Formal cooperative agreements for "mega-projects" like ITER seek to balance financial contributions with immediate benefits, such as access to technology and data.[18] A second strategy is for countries to invest in absorption of knowledge from abroad through translation facilities, exchange programs, and the like, with the goal of achieving a rough balance of trade. The United States has made fewer investments of this sort than other countries in the past, but that pattern must change in light of growing science and engineering capabilities globally.

With regard to generating long-term technology options, however, this concern is simply misplaced. The process of turning such options into viable energy resources and then bringing them to scale will take, as we have argued, a very long time. It is unlikely that any nation will reap disproportionate benefits based on its ability to control technical information over such a period. But attempts to impose such control could inhibit fruitful exchanges that might otherwise accelerate option generation. The United States ought to publicly acknowledge this logic and take steps to promote international collaboration among researchers who receive federal support.

## Forward Linkages to the Rest of the Energy Innovation System

The technical community must have substantial autonomy if it is to pursue breakthrough ideas effectively. Long-term option creation cannot be managed with the same focus on well-defined goals and milestones that should characterize the downstream stages of energy innovation. But those who might build upon breakthrough ideas and inventions downstream may have insights that would be of value to their colleagues working upstream. Input from these likely users should be sought systematically as option generation programs are formulated and projects are selected.

**Recommendation: Federally funded research aimed at creating long-term technology options should be structurally linked to the rest of the energy innovation system. DOE's Advanced Research Projects Agency–Energy (ARPA–E) can help to provide these links and deserves stable, steady funding.**

Autonomy, like cooperation, is a cherished value in the research community. Researchers argue, with some justification, that only experts who understand their proposals should decide whether they warrant support. Peer review rests on this principle. But the principle is most relevant when knowledge for its own sake is sought through the research. Research that is ultimately intended for use, even if that use is only dimly understood at the time the research is done, must be reduced to practice. In the case of energy, innovations must ultimately be made affordable. Experts from outside the research community, often from business or government, may know more about the prospects for these aspects of innovation than those generating the options.

Insights from the user community about downstream risk are likely to be particularly valuable. These risks are different from the ones that peer reviewers think about, such as the likelihood that a hypothesis will be supported. They may involve scaling up a process from the laboratory bench to the factory level. They may involve costs relative to expected alternatives. They may involve societal acceptance or integration into the broader energy system. To be sure, the uncertainty surrounding these judgments in the early stages of the innovation process is even greater than technical uncertainty. But giving them voice in the option generation stage may allow for alternative investments to be compared more realistically and for research programs to be modified to take them into account.

Decision-making in the option generation stage, therefore, ought to be balanced between peer review and user input. This concept has been gaining strength in the management of federal research programs in recent years. Reviewers of proposals to the NSF, for example, are asked to use two criteria: intellectual merit and broader impacts. The NIH Director's Council of Public Representatives, which includes patients and patients' family members, provides a formal mechanism for public input into NIH research decision making and priority-setting.

DOE's Energy Innovation Hubs could provide an institutional mechanism for linking option generation with later stages. Their goal is to bring together researchers across all sectors to tackle specific technical challenges and translate the results into practice. Of the first three hubs, which were funded in fiscal year 2010, only the one that is focusing on fuels from sunlight, based at Caltech, seeks new options. The other two, which focus on modeling and simulation for nuclear reactors and energy-efficient building systems design, seek to improve existing options.

ARPA–E may provide a more general mechanism for bringing the user's perspective to option generation. It was established in 2007 and funded initially by the 2009 stimulus package. ARPA–E's goal is to support "technologies promising genuine transformation in the ways we generate, store and utilize energy."[19] Its domain stretches from option generation to early deployment (see figure 7.1). Its program formulation and review processes engage experts from the business and investment communities as well as from science and engineering. Like the EFRCs, ARPA–E must receive regular appropriations to play an effective role in the energy innovation system. More than the EFRCs, it is likely to be subject to pressure to have an immediate impact in the marketplace, which could limit its creativity and experimental freedom. Such pressure should be resisted.

## Conclusion

The third wave of energy innovation may not seem as important as the first two. Its practical impact is decades away, and its demands on the public purse are far smaller. But even if the innovations of the first two waves are sufficient to address the challenges that are already in view,

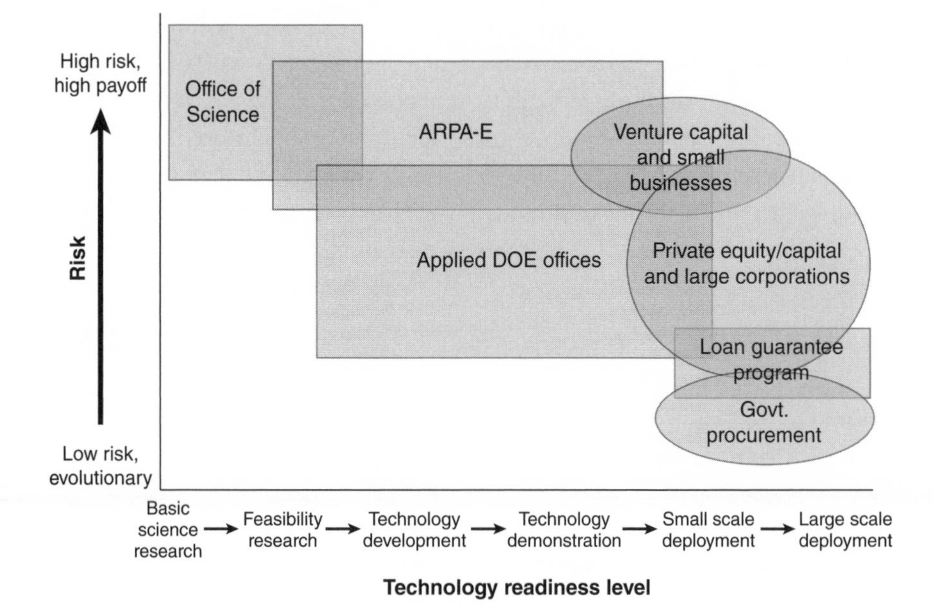


**Figure 7.1**

Energy innovation ecosystem. *Source:* Arum Majumdar, Director, ARPA–E, Woodrow Wilson International Center, Washington, DC, July 27, 2010.

there is no reason to believe that they will succeed in addressing the new challenges that may arise in the next 40 years. The grandchildren of today's children will need energy innovations too, and the options created by this generation, starting today, will greatly influence the choices available to them.

The range of possibilities is breathtaking. Contributions from the life sciences and the social sciences as well as from the physical sciences and engineering are to be expected. Which possibilities will pan out cannot be known at this point. But we can identify a number of institutions for building an energy innovation system with the potential to bring the third wave to fruition.

These include a federal system for funding the creation of new options that is much bigger than today's system, that is pluralistic in terms of sources and styles, and that fosters the open exchange of ideas, domestically and internationally. Little will be gained from trying to hoard knowledge about potential breakthrough options, while much may be gained from dialogue and collaboration. The process of creating options must be disciplined by input from participants in the downstream stages of innovation. This input provides a reality check on the enthusiasm that is an intrinsic part of the scientific and technical imagination.

The third wave of energy innovation surely will not be the last. But we will leave the challenges of subsequent waves to others, having taxed our own imaginative powers in attempting to think to the end of the current century.

# 8

# Building a New American Energy Innovation System: A Ten-Point Framework

America's innovation system is one of this country's great assets. That system must now be fully extended to meet the energy challenges of the twenty-first century. U.S. leadership in energy innovation will be essential if the world is to avoid the most harmful effects of rising greenhouse gas levels while still meeting the demand for abundant, affordable, and reliable energy. As we have emphasized in this book, nothing less than a fundamental transformation of current patterns of energy production, delivery, and use will be necessary if these goals are to be achieved.

The biggest obstacle to this transformation is that most low-carbon energy technologies available today cannot do the job. They are either too expensive or too difficult to scale, or they have other detrimental economic or environmental features. Organizational, institutional, regulatory, and behavioral challenges have impeded other options. Innovations in technology, in business models, in policies and in institutions are the means by which these barriers will be toppled.

The previous chapters have described the scale of the innovation challenge, pinpointed the weaknesses of the existing innovation system, and recommended specific actions to overcome these weaknesses. In the final chapter we draw these threads together and present an overall framework for a new system with the capacity to accelerate energy innovation across diverse fields of technology and applications over a period of decades.

We envision a far larger and much more dynamic energy innovation system than exists today, a system that will maintain the reliability and affordability of energy on which our society depends, even as it unlocks the creativity and competitive spirit of America's technical community, entrepreneurs, investors, and energy users.

This unlocking of America's energy innovation potential will not be accomplished by a centralized government program like the Manhattan Project. The task ahead is far more complex and uncertain than building an atomic bomb, even the first one. Nor can it be left to the market alone. The stimulative power of price and product competition is certainly part of what must be unlocked in some energy domains, especially in the electric power sector. But markets on their own do not always provide the right incentives for innovation, and they neglect environmental challenges like climate change.

Even if a carbon tax or cap-and-trade policy were to bring the price of energy more nearly in line with the costs that energy use actually imposes, the resulting economic incentives would not be enough to drive the energy transition on the scale and in the timeframe required. Achieving the goal of an 80 percent reduction in carbon dioxide emissions in 40 years will require accelerating all four stages of the innovation process. A carbon price will mainly affect the final stage of improvement-in-use. It will have much less impact on the first stage of the process, option creation, or on the intermediate stages of demonstration and early adoption. A healthy, well-functioning energy innovation system needs to be firing simultaneously on all four cylinders.

The coming energy transition will unfold in three successive waves of innovation. The effort to accelerate all three waves must begin right away, but the focus in each case will be on a different stage of the innovation process. The first wave, ramping up in this decade, will mainly deliver energy efficiency gains, especially in the building sector. Many useful energy efficiency technologies already exist; the challenge is to speed up their adoption on a large scale and, while this is occurring, their continued improvement—in other words, a dramatic acceleration of the fourth stage of the innovation process. The second wave will have its largest impact between 2020 and 2050, and it will yield large-scale deployment of known technologies for electricity supply and use, while driving down their costs through continual innovation. For the second wave to crest on that timescale, the immediate priority must be to accelerate the intermediate stages of the innovation process—demonstration and early adoption. A third wave of innovation, achieving scale only in the second half of the century, may result from radical technical advances generated by fundamental research in a broad range of scientific fields. Policies aimed at the third wave must focus now on accelerating new option creation.

In each of the three waves there are many candidate technologies, each with its own pros and cons and each with its own supporters. In this book we have not tried to answer the question of which innovations or which companies will or should emerge as the dominant contributors to each wave. We have instead focused on how to build an innovation system that will produce the best possible answers to this question. Building such a system will require a different kind of innovation—a rearrangement of the incentives and patterns of interaction among businesses, between business and government, within government, and between the energy industry and other sectors of the economy. We do not claim to have a complete blueprint for this new energy innovation system. In fact, we believe that attempting to build the system from such a master design would be an exercise in futility. What we offer is the first draft of a framework within which entrepreneurs, investors, government leaders, and other institution-builders will be able to create a dynamic innovation system through trial and error, conflict, and collaboration.

Ten key pieces of this framework follow.

## I. New Innovators

*At the core of the new energy innovation system will be new market entrants, both new firms and existing firms from other sectors.* The energy system is and should remain predominantly in private hands. The private firm is a remarkably flexible and innovative organizational form, especially when exposed to serious competition. No other kind of organization can scale an innovation as quickly. But the American energy industry today is dominated by large, risk-averse corporations with a history of underinvestment in innovation and, often, a strong interest in preserving the status quo. The energy industry needs an infusion of new firms, new people, and new ways of doing things. Public policy can create space for new entrants and facilitate their access to resources.

## II. Expanded Competition in Electric Power Markets

*The central front in the low-carbon energy transition will be the transformation of the electric power sector.* Expanding the domain of market competition, promoting an open industry architecture, and encouraging the entry of new competitors into newly opened segments of the electric power industry will all be powerful drivers of innovation. The most important spaces for competition and entry lie at the edges of the power grid. Independent power producers will experiment with innovative generation technologies at one end of the grid. At the other end, specialist energy service companies, demand response aggregators, and distributed generators will explore new business models, new organizational configurations, and new kinds of services for end users. To promote competition, the process of vertical disintegration of the electric utilities that began in the 1990s must be completed. The main objective at that time was lower electricity prices. Today accelerating innovation is the strongest argument for shrinking the footprint of the utilities. Expanded competition and new entrants to the power sector are the keys to all three waves of America's low-carbon energy transition.

## III. Smart Integrators

*"Smart integrator" transmission and distribution utilities, working closely with a national network of regional transmission organizations and independent system operators and with state and federal regulators, will manage the operations and development of the electric power system.* While no longer controlling the power system from end to end, the utilities that run the grid will remain the system's linchpins. They will manage the interaction of independent power producers, distributed generators, energy management service providers, customers, and many other players. Their responsibility will be to ensure that the diverse innovations arising at the edges of the grid work together to achieve the system's key objectives of lower carbon dioxide emissions, improved reliability, and greater affordability. The utilities will have to acquire new capabilities and new habits to perform this difficult task. Government regulation will still be necessary to prevent the grid's owners from exercising their monopoly power. But regulators, too, must become smarter, in tandem with the firms they are regulating. The RTOs and ISOs are an essential part of this institutional complex. They provide a transparent mechanism for operating wholesale power markets and for planning new and better transmission facilities. Congress should extend the system of regional transmission organizations to the entire country and grant RTOs and ISOs greater authority to plan and site new transmission lines.

## IV. An Invigorated Energy Efficiency Marketplace

*By the end of this decade, a thriving marketplace will speed the widespread adoption of building energy efficiency products and services.* No technological breakthroughs are necessary for the United States to become much more energy efficient. Carbon dioxide emissions can be cut significantly without impinging on the energy services (heating, lighting, etc.) that Americans rely on. In this first, efficiency-driven wave of innovation, the largest target of opportunity is in the building sector, which uses 40 percent of the nation's energy and 70 percent of the electricity. In the near term, improving building energy efficiency is the most cost-effective greenhouse gas mitigation opportunity available to the United States. The products already exist, as do the services. But they are currently confined to relatively small customer segments. The fourth stage of the innovation process, where incremental improvement and widespread adoption go hand-in-hand, has not yet taken off. Over the next decade, the most important innovations will be institutional and organizational reforms that expand the marketplace for efficiency products and services.

For new buildings and new appliances, regulations that ratchet up energy efficiency in a predictable fashion will be the key to making these markets work better. The federal government will play a key role, defining baseline standards and ensuring national compliance with them in collaboration with state and local governments (in the case of new buildings) and manufacturers (in the case of appliances). Federal testbeds (such as DOE's building energy efficiency innovation hub), along with innovative private buildings (such as those certified to a high level by the U.S. Green Building Council), will serve as the proving grounds for each ensuing iteration of these standards. These buildings will be the leading symbols of the first wave of innovation.

For retrofits of existing buildings, regulatory mandates are impractical, and a combination of financial incentives, new financing institutions, new administrative structures, and new business models will be necessary. As utilities increasingly focus on the grid integration task, new opportunities to administer building retrofit programs may appear in many states. The administrators of these programs will encourage vigorous competition in the provision of products and services, facilitate the availability of energy consumption data to third-party service providers, and sustain public information and education efforts that aim at shaping behavior. Better and more accessible information, such as an MPG-type label for all buildings, will support the deepening of efficiency product and service markets. The business model innovations of energy efficiency service providers, supported by public RD&D and information provision programs, will allow the United States to begin to close the energy efficiency gap with Europe and Japan.

## V. Regional Innovation Investment Boards

*A new group of institutions, centered on Regional Innovation Investment Boards, will unlock a second wave of innovations that will enable low-carbon electricity to supplant other energy sources.* The success of the second wave will depend on scaling up and cutting the unit costs of electricity-related services provided by low-carbon central-station power plants, distributed generation technologies, and the smart grid. Financing for demonstration and early deployment of these innovations will be mobilized and allocated in new ways. The RIIBs, membership organizations comprised of firms drawn from all segments of the electric power sector, will allocate funding to first-of-a-kind large-scale demonstration projects, "next few" post-demonstration projects, and early deployment programs. Teams proposing projects and programs will seek RIIB funding not as their sole source of finance but rather to augment their own investments and to lower their risks. In this way, RIIB investments will leverage larger amounts of private-sector funding. The RIIBs will choose among competing projects based on the strength of the proposing team, the quality of its management, and the potential of the proposal to achieve energy innovation goals, as well as the extent of self-funding. The RIIBs will compete with one another to build strong project portfolios in order to attract financial support from state-level trustee organizations. Over time each RIIB may specialize in areas of innovation of particular interest to its region.

## VI. State Energy Innovation Trustees

*Trustees set up by each state will allocate funds to the RIIBs, using the proceeds of an innovation surcharge on all retail sales of electricity within the state.* The trustees will be free to allocate their funds to RIIBs in any region. The allocation will be based on the trustee's assessment of which RIIB portfolio of demonstration and postdemonstration projects and early adoption programs most closely matches its needs. The trustee organizations will be more broadly representative of stakeholders in the electricity system than the RIIBs themselves. Their members might include a variety of business sectors, government organizations and officials, environmental and labor groups, and technical experts. All proceeds from the surcharge will have to be allocated by the trustees to the RIIBs within a year of receipt.

## VII. A Federal "Gatekeeper"

*A federal "gatekeeper" organization will certify that projects and programs presented to the RIIBs for funding have the potential to lead to significant reductions in carbon emissions at a declining unit cost over time.* To receive certification, proposals will be required to create pressure on innovators to exploit learning to reduce costs. Public subsidies for projects and programs will decline steadily on a unit basis as experience with the innovation is gained. Certification will be granted for a limited period only, and may be withdrawn if progress proves too slow. The gatekeeper will be responsible for monitoring progress. The gatekeeper will also track projects and programs targeting the same innovations to guard against duplication and overlap. However, it will take into consideration the value of pursuing several different approaches in parallel as circumstances warrant.

## VIII. Dynamic Pricing

*Some form of dynamic pricing (in which prices change during the course of the day) will make it possible for customers to make choices that stimulate innovation.* Dynamic pricing will reduce peak loads, and it will also incentivize central-station and small-scale generators as well as providers of storage and other grid services to respond to supply and demand conditions on the grid in the most effective way. Customer decision-making will also be informed by more extensive information about historical benchmarks, on-bill comparisons with neighbors, and conservation tips.

A federally sanctioned labeling program will validate the quality of meters and other equipment and help customers gain confidence that offers made by smart-grid service providers are trustworthy.

## IX. Open Grid Architecture and Customer Control

*An open architecture for distributed generation and smart grid technologies, supported by dynamic pricing, will promote innovation "behind the meter" and in the rest of the power system.* Investments in distributed generation, grid-scale storage, and smart grid technologies will help to make the power system more reliable, less wasteful, and more responsive to customer choice. In the future power grid, many entities will compete to provide many different kinds of services to the smart integrator and to customers. They will be able to "plug" into the grid and "play" their roles with minimal difficulty. Regulators and standard-setting bodies will ensure that the interfaces between service providers, users, and the grid remain open and that pricing is fully transparent. On the customer side of the meter, the architecture of the smart grid will be customer controlled rather than utility controlled. Customer-controlled architecture will give customers and third-party service providers greater freedom to experiment with devices and behaviors and provide motivation to develop innovative business models.

## X. Breakthrough Innovations

*A federal energy research structure, pluralistic in its styles, informed by user input, and larger and more diverse than today's system, will focus on the creation of new options for energy supply, delivery and use with the potential to contribute on a large scale in the second half of the century.* The Department of Energy and its large laboratories will be an important part of the system, but other government agencies, including the Department of Defense, will play a larger role than today. Coordination of the federal research effort will be led by the Executive Office of the President. The federal long-term energy research structure will foster the open exchange of ideas, both domestically and internationally, and will be linked to the downstream stages of the innovation system.

* * *

This ten-point program, as we noted previously, is only a first draft of a plan for building a new U.S. energy innovation system. But although many details need to be added, the plan already makes it clear that the goal of unlocking America's vast imaginative, entrepreneurial, managerial, and financial capabilities in the service of an accelerated energy transition will be impossible without a major institutional restructuring. We hope that others will be stimulated to add to or modify our plan. Ours is certainly not the only possible solution to this problem, and there is plenty of room for new ideas. Moreover, the task of creating a new innovation system—like the process of innovation itself—is experimental, iterative, and provisional. But any successful system will have some core attributes. It will encourage competition and the entry of outsiders, it will accelerate all four stages of the innovation process, it will ensure timely and rigorous winnowing of options at each stage, it will accommodate and exploit regional variations in innovation priorities, and it will be of a scale that matches the true magnitude of the twenty-first-century challenge.

The United States is a long way from having such a system in place today. But nothing about these requirements is inherently unattainable. Indeed, every aspect of the ideal energy innovation system has already been achieved in other fields. Energy innovation is not exactly like innovation in information technology, biotechnology, or other emerging technological areas. Change is often much slower, incumbents are more deeply entrenched, and the cost and reliability requirements for new entrants are more demanding. But we are convinced that America's energy innovation system can be renewed and greatly improved to meet the demands of this century. It is important to get started soon. There is no time to lose.

# Notes

## Chapter 1

1. National Research Council, Board on Atmospheric Sciences and Climate, Division on Earth and Life Studies, *Surface Temperature Reconstructions for the Last 2000 Years* (Washington, DC: National Academies Press, 2006), 3.

2. National Oceanic and Atmospheric Administration, *State of the Climate: Global Analysis—Annual 2010*, available at http://www.ncdc.noaa.gov/sotc/global/2010/13, accessed May 1, 2011.

3. Greenhouse gases other than $CO_2$ have added the equivalent of 70 parts per million (ppm) of $CO_2$, or about 18% of the total, to atmospheric concentrations. Other human activities such as the release of aerosols, however, have a cooling effect. The *net* warming effect of anthropogenic releases currently corresponds to the equivalent of about 380 ppm of $CO_2$. (The term "greenhouse effect" is actually a misnomer, because real greenhouses operate by preventing the convective escape of warmed air. There are no glass roofs in the earth's atmosphere.)

4. Intergovernmental Panel on Climate Change, *Fourth Assessment Report—Climate Change 2007*, available at http://www.ipcc.ch/publications_and_data/publications_and_data_reports.shtml, accessed May 3, 2011.

5. A. P. Sokolov et al., "Probabilistic Forecast for Twenty-First-Century Climate Based on Uncertainties in Emissions (Without Policy) and Climate Parameters." *J. Climate* 22 (2009): 5175–5204.

6. A recent compilation of current impacts can be found at National Aeronautics and Space Administration, *Climate Change: How Do We Know*, available at http://climate.nasa.gov/evidence, accessed October 21, 2010.

7. M. L. Parry et al., eds., *Contribution of Working Group II to the Fourth Assessment Report of the Intergovernmental Panel on Climate Change, 2007* (Cambridge: Cambridge University Press, 2007). http://www.ipcc.ch/publications_and_data/ar4/wg2/en/contents.html

8. A. P. Sokolov et al., "Probabilistic Forecast." The MIT researchers estimated that the end-of-century temperature increase relative to 1990 for the business-as-usual scenario will lie in the range from 3.81°C to 6.98°C with 90% confidence. Though not strictly comparable, the Intergovernmental Panel on Climate Change,

in its latest general assessment published in 2007, estimated an increase in the range from 2.4°C to 6.4°C (Intergovernmental Panel on Climate Change, *Fourth Assessment Report.*)

9. John P. Holdren, "The Energy/Climate-Change Challenge and the Role of Nuclear Energy in Meeting It," David J. Rose Lecture, Massachusetts Institute of Technology, 25 October 2010, video recording at http://web.mit.edu/nse/events/rose-lecture.html. We do not discuss here measures to offset or counteract the impact of increases in greenhouse gas concentrations by engineering the environment in such a way as to achieve climate modifications. Injecting sulfate aerosols into the stratosphere, enhancing the reflectivity of clouds, and building giant mirrors in space are examples of such geoengineering schemes. None of these schemes has yet been demonstrated on a significant scale, and all are likely to have negative side-effects.

10. International Energy Agency, *Key World Energy Statistics—2010* (Paris, France: International Energy Agency, 2011), 57.

11. U.S. Energy Information Administration, *State Energy Data System 2008, Consumption, Price and Expenditure Estimates* (released June 30, 2010), available at http://www.eia.gov/state/seds/

12. U.S. Energy Information Administration, "Emissions of Greenhouse Gases in the U.S.," available at http://www.eia.gov/environment/emissions/ghg_report/ghg_overview.cfm. The data for emissions during the early 1980s recession was also obtained from the Energy Information Administration website. However, the EIA has recently removed data pertaining to pre-1990 emissions from its site.

13. For a fuller account of the material in the following paragraphs, see Richard K. Lester and Ashley Finan, "Quantifying the Impact of Proposed Carbon Emission Reductions on the U.S. Energy Infrastructure," Energy Innovation Working Paper 09-006, MIT Industrial Performance Center, October 2009 (available at http://web.mit.edu/ipc/publications/pdf/09-006.pdf).

14. A target of 2% per year per capita would be approximately equal to the rate of growth achieved by the U.S. economy between 1973 and 2000, and would be well below the 2.5% growth rate achieved during the periods 1950–1973 and 1992–2000. It would, however, be an improvement on the disappointing decade of the 2000s, when annual growth averaged 1.5% even before the Great Recession of 2008–2009.

15. This calculation also assumes a future rate of growth of the U.S. population of 0.9% per year, which is consistent with both government and private projections. See, for example, J. S. Passel and D'Vera Cohn, "U.S. Population Projections: 2005-2050," Pew Research Center Report, February 2008. The U.S. Census Bureau recently projected a growth rate of 0.8% per year over this period (see http://www.census.gov/population/www/projections/usinterimproj/)

16. EPA, *Report to Congress on Server and Data Center Energy Efficiency Opportunities*, 2007; available at http://www.energystar.gov/ia/partners/prod_development/downloads/EPA_Datacenter_Report_Congress_Final1.pdf, accessed June 7, 2011.

17. U.S. Energy Information Administration, *Annual Energy Outlook 2011*, "Electricity," available at http://www.eia.gov/forecasts/aeo/MT_electric.cfm, accessed June 7, 2011.

18. T. Searchinger et al., "Use of U.S. Croplands for Biofuels Increases Greenhouse Gases Through Emissions from Land-Use Change," *Science* 319 (29 February 2008): 1238–1240.

19. Richard Newell, "Shale Gas and the Outlook for U.S. Natural Gas Markets and Global Gas Resources", presented at the Organization for Economic Cooperation and Development, Paris, France, 21 June 2011, available at http://www.eia.gov/pressroom/presentations/newell_06212011.pdf

20. U.S. Energy Information Administration, *Annual Energy Outlook 2011*, April 2011, 79, available at http://www.eia.gov/forecasts/aeo/pdf/0383(2011).pdf

21. Richard Newell, "Shale Gas, a Game Changer for U.S. and Global Gas Markets?," presented at the *Flame—European Gas Conference*, Amsterdam, Netherlands (March 2, 2010). According to the EIA the international outlook for shale gas is promising, too. A preliminary and still only partial assessment of worldwide shale gas resources indicates that they are likely to be abundant and widely dispersed. For example, sizeable shale gas endowments are believed to exist in countries like France, South Africa, and Chile that have little conventional gas reserves or production today. (See U.S. Energy Information Administration, *World Shale Gas Resources: An Initial Assessment of 14 Regions Outside the United States*, release date 5 April 2011, available at http:/www.eia.gov/analysis/studies/worldshalegas/,1.)

22. Paul Joskow, "Challenges for Creating a Comprehensive National Energy Policy," speech delivered to the National Press Club, Washington, DC, September 26, 2008.

23. Michael R. Hamilton, "Analytical Framework for Long Term Policy for Commercial Deployment and Innovation in Carbon Capture and Sequestration Technology in the United States," unpublished M.S. thesis, Technology and Policy Program, Massachusetts Institute of Technology, 2009, 33.

## Chapter 2

1. Richard K. Lester and Michael J. Piore, *Innovation—The Missing Dimension* (Cambridge, MA: Harvard University Press, 2004).

2. The consolidation of federal energy R&D activities had actually occurred two years earlier, with the formation of the short-lived Energy Research and Development Administration (ERDA) under the administration of President Gerald Ford. One of the first acts of his successor, President Jimmy Carter, was to elevate the energy portfolio to cabinet status under the leadership of the first secretary of energy, James Schlesinger.

3. Department of Energy, FY2010 Control Table by Organization, available at http://www.cfo.doe.gov/budget/10budget/Content/OrgControl.pdf

4. As a fraction of federal RD&D spending of all kinds, spending on energy RD&D has declined more or less continuously since 1978, when it accounted for 12% of the federal RD&D total. Today, even after several years of increases, it still only accounts for slightly more than 1% of total federal RD&D spending. (National Science Foundation, Table 37, "Federal research and development obligations, budget authority, and budget authority for basic research, by budget function: FY 1955–2008," available at http://www.nsf.gov/statistics/nsf07332/content.cfm?pub_id=3798&id=2, from "Federal R&D Funding by Budget Function: Fiscal Years 2006–08 Detailed Statistical Tables | NSF 07-332 | August 2007."

5. Unless otherwise stated, the data on federal energy RD&D spending provided in this section were obtained from the database maintained by Kelly Gallagher and her colleagues in the Energy Technology Innovation Policy Group at the Kennedy School of Government, Harvard University, available at http://www.belfercenter.ksg.harvard.edu/project/10/energy_technology_innovation_policy.html.

6. Linda Cohen and Roger Noll, *The Technology Pork Barrel* (Washington, DC: Brookings Institution Press, 1991).

7. National Research Council, Board on Energy and Environmental Systems, *Energy Research at DOE: Was It Worth It?* (Washington, DC: National Academies Press, 2001).

8. U.S. Energy Information Administration, Office of Coal, Nuclear, Electric and Alternative Fuels, "Federal Financial Interventions and Subsidies in Energy Markets 2007," Report number SR/CNEAF/2008-01, April 2008.

9. Details of the stimulus spending can be found at energy.gov/recovery and recovery.gov.

10. U.S. Department of Energy, Office of Energy Efficiency and Renewable Energy, Database of State Initiatives for Renewables and Efficiency, available at http://www.dsireusa.org

11. W. Nordhaus, *A Question of Balance: Weighing the Options on Global Warming Policies*, (New Haven: Yale University Press, 2008), 22.

12. R. G. Newell, A. B. Jaffe, and R. N. Stavins, "The effects of economic and policy incentives on carbon mitigation technologies," *Energy Economics* 28 (2006): 563–578.

13. International Energy Agency, *Experience Curves for Energy Technology Policy* (Paris, France: International Energy Agency, 2000). A recent comprehensive view of this subject can be found in Martin Junginger, Wilfried van Stark, and Andre Faaij, eds., *Technological Learning in the Energy Sector: Lessons for Policy, Industry, and Science* (Cheltenham, UK: Edward Elgar, 2010).

14. Along the lines of this conclusion, the MIT economist Daron Acemoglu and his colleagues showed in a recent paper that the optimal policy in an economic growth model subject to environmental constraints and limited resources involves both carbon taxes and research subsidies. They found that it would be excessively distortionary to rely solely on a tax on carbon to correct for the combined effects

of climate risks from current carbon emissions, the tendency of private actors to underinvest in R&D that could reduce future emissions, and other market distortions that tend to favor more intensive use of existing technologies over the adoption of new technologies. (Daron Acemoglu, Philippe Aghion, Leonardo Bursztyn, and David Hemous, "The Environment and Directed Technical Change," unpublished paper provided by the authors, 14 October 2009, available at http://econ-www.mit.edu/files/4670). Acemoglu and his colleagues did not, however, disaggregate the innovation process as we do here, and thus did not consider the question of which stages of the process ought to be subsidized.

15. One case study in point is David M. Hart, "Making, Breaking, and (Partially) Remaking Markets: State Regulation and Photovoltaic Electricity in New Jersey," *Energy Policy* 38 (2010): 6662–6673.

16. S. Ansolabehere et al., *The Future of Nuclear Power: An Interdisciplinary MIT Study*, Massachusetts Institute of Technology, 2003, 91.

## Chapter 3

1. We borrow this term from Peter Fox-Penner, *Smart Power* (Washington, DC: Island Press, 2010).

2. The same point might be made about ground-source heat pumps.

3. G. Pascal Zachary, *Endless Frontier: Vannevar Bush, Engineer of the American Century* (New York: Free Press, 1997).

4. W. Brian Arthur, "Positive Feedbacks in the Economy," *Scientific American* 262 (February 1990): 92–99.

5. Thomas P. Hughes, *Networks of Power* (Baltimore: Johns Hopkins University Press, 1983); Richard F. Hirsh, *Power Loss: The Origins of Deregulation and Restructuring in the American Electric Utility System* (Cambridge, MA: MIT Press, 1999).

6. Paul Joskow, "Challenges for creating a comprehensive national electricity policy," speech delivered to the National Press Club, Washington, DC, September 26, 2008, published as MIT Center for Energy and Environmental Policy Research Working Paper 08-019, September 2008.

7. Martha Derthick and Paul J. Quirk, *Politics of Deregulation* (Washington, DC: Brookings Institution Press, 1985).

8. Severin Borenstein, "The Trouble with Electricity Markets: Understanding California's Restructuring Disaster," *Journal of Economic Perspectives* 16, no. 1 (Winter 2002): 191–211. Borenstein and others also conclude that flaws in the restructuring allowed the exercise of market power, exacerbating the crisis.

9. Mason Willrich, "Electricity Transmission Policy for America: Enabling a Smart Grid, End-to-End," MIT-IPC-Energy Innovation Working Paper 09-003, July 2009.

10. Precise data on R&D spending in the electric utility industry are not easily obtained. A recent unpublished assessment of spending by the ten largest investor-

owned utilities carried out at the MIT Industrial Performance Center indicates that these firms spend about 0.2% of their revenues on R&D. Paroma Sanyal and Linda Cohen ("Powering progress: Restructuring, competition, and R&D in the U.S. electric utility industry," *Energy Journal* 30, no. 2 [2009]: 43) report a figure of $193 million for total investor-owned electric utility R&D spending in 2000 (excluding contributions to the Electric Power Research Institute). Since total revenues for major investor-owned utilities in that year were $233 billion (EIA, *Electric Power Annual 2010*, table 7.3), this implies an R&D/sales ratio of less than 0.1%.

11. The "deverticalization" of a broad range of manufacturing industries has been a focus of research at the MIT Industrial Performance Center for many years. See Suzanne Berger et al., *How We Compete* (New York: Doubleday, 2005).

12. Paul L Joskow, "Markets for Power in the United States: An Interim Assessment," *Energy Journal* 27 (2006): 4.

13. Ibid. See also William Hogan, "Electricity market structure and infrastructure," in *Acting in Time on Energy Policy*, ed. Kelly Sims Gallagher (Washington, DC: Brookings Institution Press, 2009), 128–161. We are aware that we are providing only the most basic outline of an extremely complex structure here and that important elements of it are still being perfected. In addition, there are challenges that arise from the "seams" where restructured systems meet traditional ones.

14. Fox-Penner, *Smart Power*, 189–202. The "smart integrator" and "energy services utility" do not exhaust the theoretical possibilities. Community-owned utilities specializing in medium-scale renewable resources have been an effective model, for example, in Denmark. Other countries have put ownership of their electric power industries directly in the hands of the state. France's EDF is one of the world's largest vertically integrated electric utilities, and the French state is its principal owner. However, given the difficulty of overhauling the electric sector's institutional structure, it is unlikely that the United States will adopt a foreign model. Further evolution of the two tendencies that are already visible domestically is the most likely practical path forward.

15. U.S. Energy Information Administration, *Electric Power Annual—2009*, November 2010, available at http://www.eia.gov/cneaf/electricity/epa/epa_sum.html

16. Willrich, "Electricity Transmission Policy for America," 15.

17. Paul Joskow, "Challenges for Creating a Comprehensive National Electricity Policy," 11.

18. U.S. Energy Information Administration, *Electric Power Annual,* available at http://www.eia.gov/cneaf/electricity/epa/epa_sum.html (revised April 2011), Mark Bolinger, *Surpassing Expectations: State of the U.S. Wind Power Market*, Lawrence Berkeley National Laboratory, 18 August 2009) available at http://escholarship.org/uc/item/30x3s6w0

19. This is sometimes referred to as the "missing money" problem. See Paul L. Joskow, "Competitive Electricity Markets and Investment in New Generating Ca-

pacity," MIT Center for Energy and Environmental Policy Working Paper 06-009, April 2006.

20. Environmental regulations have driven innovations in pollution control technology and fuel switching from coal to natural gas in the past. Renewable portfolio standards (RPSs) have been introduced in some states to induce regulated utilities to sign long-term power purchase agreements with owners of renewable plants.

21. Willrich, "Electricity Transmission Policy for America."

22. EnerNex Corporation, *Eastern Wind Integration and Transmission Study*, prepared for the National Renewable Energy Laboratory, revised February 2011, available at http://www.nrel.gov/wind/systemsintegration/pdfs/2010/ewits_final_report.pdf; GE Energy, *Western Wind and Solar Integration Study*, prepared for the National Renewable Energy Laboratory, May 2010, available at http://www.nrel.gov/wind/systemsintegration/wwsis.html

23. Willrich, "Electricity Transmission Policy for America," 25.

24. There are 8 NERC regions, which are divided into 130 balancing authorities, which are the basic operating units of the transmission system.

## Chapter 4

1. Unless otherwise noted, the data for this chapter are drawn from the following sources: U.S. Energy Information Administration, *Annual Energy Review 2010*, May 2010, available at http://www.eia.gov/oiaf/archive/aeo10/index.html; National Research Council, *America's Energy Future: Technology and Transformation* (Washington, DC: National Academies Press, 2010); and U.S. Department of Energy, *Buildings Energy Data Book 2009*, October 2009, available at http://buildingsdatabook.eren.doe.gov/ . This chapter also draws on the following MIT Industrial Performance Center working papers: David M. Hart, "Don't Worry About the Government? The LEED-NC 'Green Building' Rating System and Energy Efficiency in U.S. Commercial Buildings," MIT IPC Working Paper 09-001, March 2009; Richard K. Lester and Ashley Finan, "Quantifying the Impact of Proposed Carbon Emission Reductions on the U.S. Energy Infrastructure," MIT IPC Working Paper 09-006, October 2009; Jonas Nahm, "Energy Efficiency in Residential Buildings: The Case of Germany," unpublished draft of January 2011; and Rohit Sakhuja, "Energy Efficiency in Buildings," unpublished draft of March 2010.

2. Anna K. Chittum, R. Neal Elliott, and Nate Kaufman, *Trends in Industrial Energy Efficiency Programs*, American Council for an Energy-Efficient Economy Report Number IE091, September 2009, 2.

3. Much of the primary energy used to generate electricity is lost in conversion. Other losses occur in transmission and distribution.. The 70% figure in the text refers to the electricity that is actually delivered to end-users by the electric grid.

4. The main source of energy use data for buildings, the quadrennial Commercial Buildings Energy Consumption Survey (CBECS), did not yield usable data in

2007, and the 2011 CBECS was suspended in April 2011 as a result of cuts in the EIA budget imposed by the FY2011 continuing resolution.

5. Bruce Hunn, interview by David Hart, August 5, 2008.

6. National Research Council, *America's Energy Future.*

7. Steve Selkowitz, "Zero Energy Buildings," presentation at MIT, November 18, 2008.

8. California Energy Commission, *2005 Building Energy Efficiency Standards: Residential Compliance Manual*, CEC-400-2005-005-CMF (rev. 3), March 2005, 1–4.

9. Devorah Eden, "Energy Efficiency Actions and Title 24 Changes," April 10, 2009, p. 4, available at http://www.energy.ca.gov/renewables/06-NSHP-1/documents/2009-04-10_workshop/presentations/Eden-Energy_Efficiency_and_2008_Standards.pdf, accessed June 28, 2011.

10. Arthur H. Rosenfeld with Deborah Poskanzer, "A Graph Is Worth a Thousand Gigawatt-Hours: How California Came to Lead the United States in Energy Efficiency," *Innovations* 4, no. 4 (Fall 2009): 57–80.

11. National Research Council, *America's Energy Future,* 42.

12. MIT Energy Innovation Project, Energy Efficiency Workshop: Commercial Buildings Retrofits, Cambridge, MA, March 5, 2009.

13. Ibid.

14. American Council for an Energy-Efficient Economy, *State Energy Efficiency Scorecard 2010*, report E107, October 13, 2010.

15. Office of the Deputy Undersecretary of Defense, *Annual Energy Management Report—Fiscal Year 2009*, May 2010, available at http://www.acq.osd.mil/ie/energy/library/aemr_fy_09_may_2010.pdf

16. We will use the term "new buildings" in the rest of this chapter but our intention is to encompass major renovations within it as well.

17. Architecture 2030, "A Historic Opportunity," http://architecture2030.org/the_solution/buildings_solution_how, accessed May 17, 2011.

18. D. L. Shankle, J. A. Merrick, and T. L. Gilbride, "A History of the Building Energy Standards Program," PNL-9386, Pacific Northwest Laboratory, February 1994.

19. DOE at the time of this writing was in the process of certifying the stringency of ASHRAE 90.1-2010.

20. Some "home rule" states, in which localities have strong prerogatives relative to the state, are forbidden from adopting such a code.

21. The current status of state code adoption is tracked by DOE's Pacific Northwest Laboratory at www.energycodes.gov/states.

22. California Energy Commission, *AB 2160 Green Building Report*, CEC-400-2008-005-CMF, January 2008.

23. See, for example, Amanda Ricker, "City to lay off code enforcement officer," *Bozeman Daily Chronicle*, June 2, 2011.

24. In late 2010 DOE awarded $7 million to 24 states for activities related to the adoption of and compliance with the most current building energy codes. DOE, Building Energy Codes Program, "Energy Efficiency Contracts Awarded to 24 States," http://www.energycodes.gov/arra/state_funding_awards.stm, accessed May 19, 2011.

25. This approach builds on provisions in the American Recovery and Reinvestment Act (ARRA) of 2009, although the funds were allocated as rapidly as possible in that case in order to stimulate the economy. ARRA requires states to adopt the most recent codes and achieve 90% compliance within eight years.

26. EDAW, Inc., "Energy Code Case Study—Title 24" (Seattle: Seattle New Building Energy Efficiency Policy Analysis), November 4, 2008, http://www.seattle.gov/environment/documents/GBTF_NewBldg_Title24_Case_Study.pdf. The Passive House Institute-U.S. estimates that to meet the extremely stringent Passivhaus standard imposes a premium of about 10% in the U.S. and less than 5% in Europe. See Tom Zeller, Jr., "Can We Build in a Brighter Shade of Green?," *New York Times*, September 25, 2010.

27. Kenneth Gillingham, Richard Newell, and Karen Palmer, "Energy Efficiency Policies: A Retrospective Examination," *Annual Review of Environment and Resources* 31 (2006): 182.

28. National Research Council, *Real Prospects for Energy Efficiency in the United States* (Washington, DC: National Academies Press, 2010), 44.

29. National Research Council, *America's Energy Future,* 192.

30. See, for example, Gillingham, "Energy Efficiency Policies"; Stephen Meyers, James McMahon, and Michael McNeil, *Realized and prospective impacts of U.S. energy efficiency standards for residential appliances,* Lawrence Berkeley National Laboratory report LBNL-56417, 2005; Stephen Nadel, *Annual Review of Energy and Environment* 27 (2002):159–192; Richard G. Newell, Adam B. Jaffe, and Robert N. Stavins, "The Induced Innovation Hypothesis and Energy-Saving Technological Change," *Quarterly Journal of Economics* 114, no. 3 (August 1999): 941–975.

31. U.S. Government Accountability Office, "Long-standing Problems with DOE's Program for Setting Efficiency Standards Continue to Result in Forgone Energy Savings," report 07-42, January 2007. The historical roles of Congress and the president have been reversed by recent elections, with Congress pulling back the reins while the administration seeks to push forward aggressively.

32. Max Neubauer et al., "Ka-BOOM! The Power of Appliance Standards," American Council for an Energy-Efficient Economy (ACEEE), Report Number ASAP-7/ACEEE-A091, July 2009, vi.

33. Ibid.

34. Mitchell Rosenberg and Lynn Hoefgen, "Market Effects and Market Transformation: Their Role in Energy Efficiency Program Design and Evaluation," California Institute for Energy and Environment, March 2009, 18.

35. American Council for an Energy-Efficient Economy, "Major Home Appliance Efficiency Gains to Deliver Huge National Energy and Water Savings and Help to Jump Start the Smart Grid," press release, August 3, 2010.

36. The reader will recall that major renovations are treated the same way as new buildings. Control systems will be considered as part of the "smart grid" in chapter 6.

37. Rachel Gold and Steven Nadel, "Energy Efficiency Tax Incentives, 2005-2011: How Have They Performed," ACEEE White Paper, June 2011.

38. Consortium for Energy Efficiency (CEE), "The State of the Efficiency Program Industry: 2009 Expenditures, Impacts, 2010 Budgets," Boston, MA, December 2010, pp. 16 and 24, available at http://www.cee1.org/files/2010%20State%20of%20the%20Efficiency%20Program%20Industry.pdf

39. Katherine Friedrich et al., "Saving Energy Cost-Effectively: A National Review of the Cost of Energy Saved Through Utility-Sector Energy Efficiency Programs," Report Number U092, American Council for an Energy-Efficient Economy, September 2009, available at http://www.aceee.org/sites/default/files/publications/researchreports/U092.pdf, Toshi H. Arimura, et al., "Cost-Effectiveness of Electricity Energy Efficiency Programs," Resources for the Future Report DP 09-48-REV, revised April 2011.

40. Friedrich, "Saving Energy Cost-Effectively," 10.

41. John Podesta and Karen Kornbluh, "The Green Bank," Center for American Progress, May 2009, http://www.americanprogress.org/issues/2009/05/pdf/green_bank_memo.pdf.

42. The term "clean energy" in this context is, in our view, unfortunate. The PACE model has been more avidly promoted for small-scale renewables, which may or may not be cost-effective, than for energy efficiency, which is far more likely to be.

43. Martin Kushler, Dan York, and Patti Witte, *Meeting Aggressive New State Goals for Utility-Sector Energy Efficiency*, ACEEE, March 2009.

44. Some states have begun to impose targets for energy efficiency gains, sometimes known as energy efficiency portfolio standards. Such targets may have a symbolic value, but they are likely to be superfluous for the effective smart integrator and may even drive poor investment decision-making in some circumstances.

45. Regulations on access to utility information on energy use vary from state to state.

46. Barry B. LePatner, *Broken Buildings, Busted Budgets* (Chicago: University of Chicago Press, 2007).

47. Marilyn Brown, "Energy-efficient buildings: does the marketplace work?" Oak Ridge National Laboratory, 1997.

48. Brian Wright and Tiffany Shih, "Agricultural Innovation," in *Accelerating Energy Innovation: Lessons from Multiple Sectors,* ed. Rebecca M. Henderson and Richard G. Newell (Chicago: University of Chicago Press, 2011).

49. Rosenfeld, "A Graph Is Worth a Thousand Gigawatt-Hours," 65.

50. American Physical Society, *Energy Future: Think Efficiency*, September 16, 2008, Executive Summary available at http://www.aps.org/energyefficiencyreport/summary/energyexecsum.pdf.

51. Government Accountability Office, "Agencies Are Taking Steps to Meet High-Performance Federal Building Requirements, but Face Challenges," report GAO-10-22, October 2009.

52. U.S. Navy, Energy, Environment, and Climate Change, "Energy," http://greenfleet.dodlive.mil/home/energy, accessed June 28, 2011.

## Chapter 5

1. MIT Energy Initiative, *The Future of Natural Gas: An Interdisciplinary MIT Study*, 2011, 69, available at http://web.mit.edu/mitei/research/studies/index.shtml. As we discussed in chapter 1, substitution of gas for coal has the potential to reduce carbon emissions per unit of electricity output by 50% or more. We endorse using electricity from new as well as existing gas-fired power plants to substitute for old coal plants in the next decade or two.

2. This colorful but imprecise term invites confusion with another commonly cited "valley of death," where the latter refers to the shortage of seed capital needed to translate the results of laboratory research into a sufficiently convincing business concept to attract angel or early-stage venture capital investors.

3. Shikhar Ghosh and Ramana Nanda, "Venture Capital Investment in the Clean Energy Sector," MIT Industrial Performance Center Working Paper, August 1, 2010.

4. Ibid., 8.

5. Ashley Finan, unpublished Ph.D. thesis, Department of Nuclear Science and Engineering, Massachusetts Institute of Technology, forthcoming 2011. As Finan notes, the legacy of the Shippingport project has been much debated. The project had initially been planned as one of several parallel government-funded demonstrations of alternative nuclear power reactor technologies. But federal budget constraints, political conflicts over the proper roles for government and private industry in the power sector, and the greater maturity of pressurized-water reactor (PWR) technology thanks to its already-extensive role in naval propulsion left Shippingport as the only major government-funded nuclear reactor demonstration. The success of the project, which was closely supervised by Admiral Hyman Rickover, helped establish an unassailable lead for PWR technology for commercial power applications.

6. See, for example, Linda Cohen and Roger Noll, *The Technology Pork Barrel* (Washington, DC: Brookings Institution Press, 1991).

7. John M. Deutch, "An Energy Technology Corporation will Improve the Federal Government's Efforts to Accelerate Energy Innovation," Hamilton Project Discussion Paper 2011-05, Brookings Institution, Washington, DC, May 2011.

8. Jeff Bingaman, "An Energy Agenda for the Next Congress," *Issues in Science and Technology*, Spring 2011, 35–42.

9. John Podesta and Karen Kornbluh, “The Green Bank,” Center for American Progress, May 21, 2009, available at http://www.americanprogress.org/issues/2009/05/green_bank.html.

10. John Deutch, “An Energy Technology Corporation.” Somewhat similar proposals have been implemented elsewhere. Sustainable Development Technology Canada, a non-profit foundation, operates two funds totaling about $1 billion designed to fill the financing gap between early stage research and commercial deployment. One of the funds is focused on financing first-of-a-kind large-scale demonstration facilities for next-generation renewable fuels. The organization was created and is funded by the Canadian government and is accountable to Parliament. See http://www.sdtc.ca/index.php?page=home&hl=en_CA

11. American Energy Innovation Council, *A Business Plan for America’s Energy Future,* available at http://www.americanenergyinnovation.org/

12. Some of the ideas on which this proposal is based were originally suggested by Paul Romer, “Implementing a National Technology Strategy with Self-Organizing Industry Investment Boards,” *Brookings Papers: Microeconomics* 2 (1993): 345–390.

13. Today EPRI receives just $180 million per year in membership revenues, a tiny fraction (one-twentieth of one percent) of electric power industry sales—and not nearly enough to fund a serious program of technology demonstrations. Moreover, most of the funds are used to support near-term engineering work aimed at improvements in operations and maintenance.

14. Each NERC regional coordinating council is comprised of all the investor-owned utilities, public power authorities, independent power producers, and large energy users in the region. So there would be a strong overlap between the membership of the NERC regions and the membership of the RIIBs. But the two functions—system reliability coordination and long-term innovation investment—are very different, so there might not be much benefit to combining them in a single organization. A possible additional complication is that three of the NERC regions, the Western Electricity Coordinating Council, the Midwest Reliability Organization, and the Northeast Power Coordinating Council, extend into Canada, and two others, Texas and Florida, are single-state regions.

15. President’s Council of Advisors on Science and Technology (PCAST), *Report to the President on Accelerating the Pace of Change in Energy Technologies Through an Integrated Federal Energy Policy*, Executive Office of the President, November 2010, available at http://www.whitehouse.gov/sites/default/files/microsites/ostp/pcast-energy-tech-report.pdf, p. viii. PCAST proposes that $10 billion in energy RD&D funding (out of a proposed expansion of $16 billion) be derived from new funding streams.

16. Excluding Recovery Act (stimulus) funding the DOE laboratories in 2010 accounted for 75% of total DOE spending on science and 45% of spending on applied research, development and demonstration (John M. Deutch, “An Energy Technology Corporation,” 13). The laboratories’ efforts in energy technology commercialization have long been criticized, although there is evidence that this aspect of their performance has improved. (See Adam Jaffe and Josh Lerner,

"Reinventing public R&D: Patent policy and the commercialization of national laboratory technologies," *RAND Journal of Economics* 32, no.1 (Spring 2001): 167–198).

## Chapter 6

1. This chapter draws on four MIT working papers: Neil Peretz, "Standards in Information Technology and the Smart Grid—Are Historical Analogies Useful?", MIT Industrial Performance Center Working Paper 11-002, July 2011; David M. Hart, "Making, Breaking, and (Partially) Remaking Markets: State Regulation and Photovoltaic Electricity in New Jersey," *Energy Policy* 38 (2010): 6662–6673 (originally published as MIT Industrial Performance Center Working Paper 09-005, July 2009); Harvey Michaels and Kat Donnelly, "Enabling Energy Efficiency on the Customer Side of the Smart Grid," unpublished working paper, November 2009; and Shikhar Ghosh and Ramana Nanda, "Venture Capital Investment in the Clean Energy Sector," MIT Industrial Performance Center Working Paper 10-004, August 2010.

2. One "grand plan" for solar power envisions the batteries in electric cars providing large-scale storage. See Ken Zweibel, James Mason and Vasilis Fthenakis, "A Solar Grand Plan," *Scientific American*, January 2008.

3. U.S. Department of Energy, Office of Energy Efficiency and Renewable Energy, *Database of State Incentives for Renewables and Efficiency*, www.dsireusa.org, accessed June 16, 2011.

4. The Network for New Energy Choices's annual *Freeing the Grid* (New York, 2007–2010) report grades every state on its interconnection and net metering policies. The December 2010 version, p. 13, shows the time trend for grades over a four-year period.

5. Felicity Barringer, "California wants to control home thermostats," *New York Times*, January 1, 2008.

6. Michaels and Donnelly, "Enabling Energy Efficiency," 14.

7. Ahmad Faruqui, "Dynamic Pricing—the Top Ten Myths," Brattle Group, April 7, 2011, available at http://www.brattle.com/_documents/UploadLibrary/Upload936.pdf, accessed June 18, 2011.

8. John Wellinghoff presentation at MIT, October 30, 2009.

9. About 0.3% of electricity sales in 2009 were made through "green power" programs in competitive or vertically integrated markets. Lori Bird and Jenny Sumner, *Green Power Marketing in the United States: A Status Report (2009 Data)*, National Renewable Energy Laboratory, Technical Report NREL/TP-6A20-49403, September 2010.

10. If a utility-controlled architecture is installed, then the smarter smart meter would also presumably be paid for through a rate-setting proceeding.

11. Faruqui, "Dynamic Pricing."

12. Aaron Smith, *Home Broadband 2010*, Pew Internet and American Life Project, August 2010, available at http://www.pewinternet.org/Reports/2010/Home-Broadband-2010.aspx, accessed June 19, 2011.

13. Federal Communications Commission, *National Broadband Plan*, April 2010, chap. 14, http://www.broadband.gov/issues/energy-and-the-environment.html, accessed June 19, 2011.

14. Electric Power Research Institute, *Estimating the Costs and Benefits of the Smart Grid*, Report Number 1022519, Palo Alto, CA, March 2011.

15. Of the total cost for the smart grid build-out estimated by EPRI, roughly 8.5% would be spent at customer premises, 70% on the distribution system, and 21.5% on the transmission system.

16. "Maryland rejects Baltimore Gas and Electric's Smart Grid Plan," *Smart Grid Legal News,* available at http://www.smartgridlegalnews.com/regulatory-concerns-1/maryland-will-not-be-the-betamax-of-the-smart-grid/ (accessed July 6, 2001.) In denying the proposal, the Maryland Public Service Commission commented that "the field of modern technology is replete with examples of innovations once considered the leaders into a new era that were never widely adopted. All the federal funding in the world would not have made Sony's Betamax a wise investment, for example. Those who invest in new technology as it becomes available often find themselves re-investing much sooner than they anticipated."

17. This problem would be less pronounced under the utility-controlled smart grid architecture, but it would not disappear because customers and third-party service providers would still retain substantial discretion.

18. We noted earlier in this chapter that RIIB investments could also encompass distributed generation, storage, and the customer-side smart grid.

19. Michaels and Donnelly, "Enabling Energy Efficiency," 5–6.

20. Martin Kenney, "Venture Capital Investments in the Greentech Industries," in *Handbook of Research on Energy Entrepreneurship*, ed. Rolf Wustenhagen and Robert Wuebker (Northampton, MA: Edward Elgar, 2011), 214–228.

21. The 2009 federal stimulus package (ARRA) funded a Smart Grid Demonstration Program within the Department of Energy as well as a much larger program of Smart Grid Investment Grants that are intended to support immediate commercial deployment. Details of these awards can be found at http://www.smartgrid.gov/recovery_act, accessed June 24, 2011.

22. National Science Foundation Workshop on the Future Power Engineering Workforce (held November 29–30, 2007), Final Report, September 5, 2008, available at http://ecpe.ece.iastate.edu/nsfws/Report%20of%20NSF%20Workshop.pdf (accessed on July 6, 2011).

23. Dan Charles, "Students Energized by Power Engineering," *Science* 324 (10 April 2009): 175.

## Chapter 7

1. U.S. Department of Energy, Advanced Research Projects Agency-Energy, http://arpa-e.energy.gov/ProgramsProjects/Electrofuels.aspx (accessed on July 6, 2011).

2. Dan E. Robertson et al., "A New Dawn for Industrial Photosynthesis," *Photosynthesis Research,* published online, 13 February 2011, available at http://www.springerlink.com/content/j1414q2u5w25h788/fulltext.html (accessed on 7 July 2011).

3. U.S. Energy Information Administration, *Alternatives to Traditional Transportation Fuels*, release date April 28, 2011, available at http://www.eia.gov/renewable/alternative_transport_vehicles/index.cfm, table C1 (accessed on July 16, 2011).

4. Further information on MIT's SENSEable City Laboratory is available at http://senseable.mit.edu/. Results from the concrete sensing skin research project were reported in Matthias Kollosche, Hristiyan Stoyanov, Simon Laflamme, and Guggi Kofod, "Strongly Enhanced Sensitivity in Elastic Capacitive Strain Sensors," *Journal of Materials Chemistry* 21 (2011), 8292–8294. The Concrete Sustainability Hub (at http://web.mit.edu/cshub/research/index.html) is based in MIT's Department of Civil and Environmental Engineering. The Changing Places and House_n Research Consortia are described at http://web.media.mit.edu/~kll/A_CP%20Research%20Topics.pdf.

5. National Science Foundation, *Science and Engineering Indicators 2010* (Washington, DC, 2010), chapter 4, sidebar 2, http://www.nsf.gov/statistics/seind10/c4/c4s.htm#sb2, accessed July 1, 2011.

6. International Energy Agency, *Global Gaps in Clean Energy Research, Development, and Demonstration* Paris, December 2009, 2.

7. U.S. Energy Information Administration, *The Financial Reporting System*, available at http://www.eia.gov/emeu/finance/page1a.html.

8. American Association for the Advancement of Science, "Historical Data on Federal R&D, FY1976-2009," available at http://www.aaas.org/spp/rd/tbbas09p.pdf, accessed July 1, 2011.

9. Paul Berg and others, Letter to President Obama, July 16, 2009, http://www.fas.org/press/news/2009/july_nobelist_letter_to_obama.html.

10. American Energy Innovation Council, *A Business Plan for America's Energy Future*, June 2010, available at http://www.americanenergyinnovation.org/full-report.

11. President's Council of Advisors on Science and Technology, *Report to the President on Accelerating the Pace of Change in Energy Technologies Through an Integrated Federal Energy Policy* (Washington, DC, November 2010).

12. Richard K. Lester and Michael J. Piore, *Innovation—The Missing Dimension* (Cambridge, MA: Harvard University Press, 2004).

13. Consortium for Science, Policy, and Outcomes (Arizona State University) and Clean Air Task Force, *Innovation Policy for Climate Change: A Report to the Nation*, September 2009.

14. National Science Board, *Building a Sustainable Energy Future: U.S. Actions for an Effective Energy Economy Transformation*, NSB-09-55, August 2009.

15. James Duderstadt et al., *Energy Discovery-Innovation Institutes: A Step Toward America's Energy Sustainability*, Brookings Institution, February 2009.

16. National Science Board, *Science and Engineering Indicators 2010*, Washington, 2010, chapter 5, section 4, available at http://www.nsf.gov/statistics/seind10/c5/c5s4.htm#s2

17. Ryan Zelnio, "The Global Structure of International Scientific Collaborations," paper presented to the 13th International Society of Scientometrics and Informetrics Conference, Durban, South Africa, July 2011. The energy field is defined in this paper by Thomson Reuters' Web of Knowledge database.

18. Unfortunately, disputes over which nations will receive which benefits have contributed to the delays that have plagued the ITER project.

19. U.S. Department of Energy, Advanced Research Projects Agency–Energy, "Mission," http://arpa-e.energy.gov/About/Mission.aspx, accessed July 4, 2011.

# Bibliography


Acemoglu, Daron, Philippe Aghion, Leonardo Bursztyn, and David Hemous. "The Environment and Directed Technical Change." Unpublished paper provided by the authors, 14 October 2009. Available at http://econ-www.mit.edu/files/4670.

American Association for the Advancement of Science. "Historical Data on Federal R&D, FY1976–2009." Available at http://www.aaas.org/spp/rd/tbbas09p.pdf.

American Council for an Energy-Efficient Economy. "Major Home Appliance Efficiency Gains to Deliver Huge National Energy and Water Savings and Help to Jump Start the Smart Grid." Press release, August 3, 2010. Washington, DC: ACEEE, 2010.

American Council for an Energy-Efficient Economy. *State Energy Efficiency Scorecard 2010*. Report E107. Washington, DC: ACEEE, 2010.

American Energy Innovation Council. *A Business Plan for America's Energy Future.* 2010 Available at http://www.americanenergyinnovation.org/

American Physical Society. *Energy Future: Think Efficiency*. Washington, DC: APS, 2008.

Ansolabehere, S., et al. *The Future of Nuclear Power: An Interdisciplinary MIT Study.* Cambridge, MA: Massachusetts Institute of Technology, 2003. Architecture 2030. "A Historic Opportunity." Available at http://architecture2030.org/the_solution/buildings_solution_how.

Arimura, Toshi H., et al. *Cost-Effectiveness of Electricity Energy Efficiency Programs*. Resources for the Future Report, DP 09-48-REV, April 2011.

Arthur, W. "Brian. "Positive Feedbacks in the Economy." *Scientific American* 262 (February 1990): 92–99.

Barringer, Felicity. "California Wants to Control Home Thermostats." *New York Times*, January 1, 2008.

Berg, Paul, et al. "Letter to President Obama." July 16, 2009. Available at http://www.fas.org/press/news/2009/july_nobelist_letter_to_obama.html.

Berger, Suzanne, et al. *How We Compete*. New York: Doubleday, 2005.

Bingaman, Jeff. "An Energy Agenda for the Next Congress." *Issues in Science and Technology* 27 (3) (2011): 35–42.

Bird, Lori, and Jenny Sumner. *Green Power Marketing in the United States: A Status Report (2009 Data)*. Technical Report NREL/TP-6A20-49403. Golden, CO: National Renewable Energy Laboratory, 2010.

Bolinger, Mark. *Surpassing Expectations: State of the U.S. Wind Power Market.* Berkeley: Lawrence Berkeley National Laboratory, 2009.

Borenstein, Severin. "The Trouble with Electricity Markets: Understanding California's Restructuring Disaster." *Journal of Economic Perspectives* 16 (1) (Winter 2002): 191–211.

Brown, Marilyn. *Energy-Efficient Buildings: Does the Marketplace Work?* Oak Ridge, TN: Oak Ridge National Laboratory, 1997.

California Energy Commission. *2005 Building Energy Efficiency Standards: Residential Compliance Manual*. CEC-400-2005-005-CMF rev. 3, March 2005.

California Energy Commission. *AB 2160 Green Building Report*. Report CEC-400-2008-005-CMF, January 2008.

Charles, Dan. "Students Energized by Power Engineering." *Science* 324 (April 10, 2009): 175.

Chittum, Anna K. R., Neal Elliott, and Nate Kaufman. *Trends in Industrial Energy Efficiency Programs*. American Council for an Energy-Efficient Economy Report Number IE091. Washington, DC: ACEEE, 2009.

Cohen, Linda, and Roger Noll. *The Technology Pork Barrel*. Washington, DC: Brookings Institution Press, 1991.

Consortium for Energy Efficiency (CEE). *The State of the Efficiency Program Industry: 2009 Expenditures, Impacts, 2010 Budgets*. Boston: CEE, 2010.

Consortium for Science, Policy, and Outcomes (Arizona State University) and Clean Air Task Force. *Innovation Policy for Climate Change: A Report to the Nation*. Washington, DC: CSPO, 2009.

Derthick, Martha, and Paul J. Quirk. *Politics of Deregulation*. Washington, DC: Brookings Institution Press, 1985.

Deutch, John M. *An Energy Technology Corporation will Improve the Federal Government's Efforts to Accelerate Energy Innovation*. Hamilton Project Discussion Paper 2011-05. Washington, D,C,: Brookings Institution, 2011.

Duderstadt, James, et al. *Energy Discovery-Innovation Institutes: A Step Toward America's Energy Sustainability*. Washington: Brookings Institution, 2009.

EDAW, Inc. "Energy Code Case Study—Title 24." Seattle: Seattle New Building Energy Efficiency Policy Analysis, 2008. Available at http://www.seattle.gov/environment/documents/GBTF_NewBldg_Title24_Case_Study.pdf.

Eden, Devorah. "Energy Efficiency Actions and Title 24 Changes." California Energy Commission, April 10, 2009. Available at http://www.energy.ca.gov/renewables/06-NSHP-1/documents/2009-04-10_workshop/presentations/Eden-Energy_Efficiency_and_2008_Standards.pdf.

Electric Power Research Institute. *Estimating the Costs and Benefits of the Smart Grid*. Report Number 1022519. Palo Alto, CA: EPRI, 2011.

EnerNex Corporation. *Eastern Wind Integration and Transmission Study*. Golden, CO: National Renewable Energy Laboratory, 2011.

Faruqui, Ahmad. "Dynamic Pricing—the Top Ten Myths." Washington, D,C,: Brattle Group, April 7, 2011. Available at http://www.brattle.com/_documents/UploadLibrary/Upload936.pdf.

Finan, Ashley. Unpublished Ph.D. Thesis, Department of Nuclear Science and Engineering, Massachusetts Institute of Technology, forthcoming 2011.

Fox-Penner, Peter. *Smart Power*. Washington, D,C,: Island Press, 2010.

Friedrich, Katherine, et al. *Saving Energy Cost-Effectively: A National Review of the Cost of Energy Saved Through Utility-Sector Energy Efficiency Programs*. Report Number U092, Washington, DC: ACEEE, 2009.

Gallagher, Kelly, et al. *DOE Budget Authority for Energy Research, Development, & Demonstration Database*. Available at http://belfercenter.ksg.harvard.edu/files/doe_energy-tech-spending1978-fy12R_march3f.xls.

G. E. Energy. *Western Wind and Solar Integration Study*. Golden, CO: National Renewable Energy Laboratory, 2011.

Ghosh, Shikhar, and Ramana Nanda. "Venture Capital Investment in the Clean Energy Sector." MIT IPC Working Paper 10-004.

Gillingham, Kenneth, Richard Newell, and Karen Palmer. "Energy Efficiency Policies: A Retrospective Examination." *Annual Review of Environment and Resources* 31 (2006): 161–192.

Gold, Rachel, and Steven Nadel. "Energy Efficiency Tax Incentives, 2005-2011: How Have They Performed." ACEEE White Paper. Washington, DC: ACEEE, 2011.

Hamilton, Michael R. "Analytical Framework for Long Term Policy for Commercial Deployment and Innovation in Carbon Capture and Sequestration Technology in the United States." Unpublished M.S. thesis, Technology and Policy Program, Massachusetts Institute of Technology, 2009.

Hart, David M. "Don't Worry About the Government? The LEED-NC "Green Building" Rating System and Energy Efficiency in U.S. Commercial Buildings." MIT IPC Working Paper 09-001, March 2009.

Hart, David M. "Making, Breaking, and (Partially) Remaking Markets: State Regulation and Photovoltaic Electricity in New Jersey." *Energy Policy* 38 (2010): 6662–6673.

Hirsh, Richard F. *Power Loss: The Origins of Deregulation and Restructuring in the American Electric Utility System*. Cambridge, MA: MIT Press, 1999.

Hogan, William. Electricity Market Structure and Infrastructure. In *Acting in Time on Energy Policy*, ed. Kelly Sims Gallagher, 128–161. Washington, DC: Brookings Institution Press, 2009.

Holdren, John P. "The Energy/Climate-Change Challenge and the Role of Nuclear Energy in Meeting It." David J. Rose Lecture, Massachusetts Institute of Technology, 25 October 2010. Available at http://web.mit.edu/nse/events/rose-lecture.html.

Hughes, Thomas P. *Networks of Power*. Baltimore: Johns Hopkins University Press, 1983.

Hunn, Bruce. Interview (telephone) by David Hart, August 5, 2008.

Intergovernmental Panel on Climate Change. *Fourth Assessment Report—Climate Change 2007*. Geneva: IPCC, 2007.

International Energy Agency. *Experience Curves for Energy Technology Policy*. Paris, France: International Energy Agency, 2000.

International Energy Agency. *Global Gaps in Clean Energy Research, Development, and Demonstration*. Paris: IEA, 2009.

International Energy Agency. *Key World Energy Statistics*. Paris: International Energy Agency, 2011.

Jaffe, Adam, and Josh Lerner. "Reinventing Public R&D: Patent Policy and the Commercialization of National Laboratory Technologies." *Rand Journal of Economics* 32 (1) (Spring 2001): 167–198.

Joskow, Paul L. "Markets for Power in the United States: An Interim Assessment." *Energy Journal* 27 (2006): 1–36.

Joskow, Paul L. "Competitive Electricity Markets and Investment in New Generating Capacity." MIT Center for Energy and Environmental Policy Working Paper 06-009, April 2006.

Joskow, Paul. "Challenges for Creating A Comprehensive National Electricity Policy." MIT Center for Energy and Environmental Policy Research Working Paper 08-019. Cambridge, MA: MIT, 2008.

Junginger, Martin, Wilfried van Stark, and Andre Faaij, eds. *Technological Learning in the Energy Sector: Lessons for Policy, Industry, and Science*. Cheltenham, UK: Edward Elgar, 2010.

Kenney, Martin. Venture Capital Investments in the Greentech Industries. In *Handbook of Research on Energy Entrepreneurship*, ed. Rolf Wustenhagen and Robert Wuebker, 214–228. Northampton, MA: Edward Elgar, 2011.

Kollosche, Matthias, Hristiyan Stoyanov, Simon Laflamme, and Guggi Kofod. "Strongly Enhanced Sensitivity in Elastic Capacitive Strain Sensors." *Journal of Materials Chemistry* 21 (2011): 8292–8294.

Kushler, Martin, Dan York, and Patti Witte. *Meeting Aggressive New State Goals for Utility-Sector Energy Efficiency*. Washington, DC: ACEEE, 2009.

LePatner, Barry B. *Broken Buildings, Busted Budgets*. Chicago: University of Chicago Press, 2007.

Lester, Richard K., and Ashley Finan. "Quantifying the Impact of Proposed Carbon Emission Reductions on the U.S. Energy Infrastructure." MIT Industrial Performance Center: Energy Innovation Working Paper 09-006, 2009.

Lester, Richard K., and Michael J. Piore. *Innovation—The Missing Dimension*. Cambridge, MA: Harvard University Press, 2004.

Meyers, Stephen, James McMahon, and Michael McNeil. *Realized and Prospective Impacts of U.S. Energy Efficiency Standards for Residential Appliances*. Lawrence Berkeley National Laboratory report LBNL-56417, 2005.

Michaels, Harvey, and Kat Donnelly. "Enabling Energy Efficiency on the Customer Side of the Smart Grid." Unpublished MIT IPC Working Paper, November 2009.

MIT Energy Initiative. *The Future of Natural Gas: An Interdisciplinary MIT Study*. Cambridge, MA: MIT, 2011.

MIT Energy Initiative. *The Future of Nuclear Fuel Cycle*. Cambridge, MA: MIT, 2010.

MIT Energy Innovation Project. "Energy Efficiency Workshop: Commercial Buildings Retrofits." Cambridge, MA, March 5, 2009.

Nadel, Stephen. "Appliances and Energy Equipment Standards." *Annual Review of Energy and the Environment* 27 (2002): 159–192.

Nahm, Jonas. "Energy Efficiency in Residential Buildings: The Case of Germany." Unpublished MIT IPC Working Paper, January 2011.

National Research Council, Board on Atmospheric Sciences and Climate, Division on Earth and Life Studies. *Surface Temperature Reconstructions for the Last 2000 Years*. Washington, DC: National Academies Press, 2006.

National Research Council, Board on Energy and Environmental Systems. *Energy Research at DOE: Was It Worth It?* Washington, DC: National Academies Press, 2001.

National Research Council. *America's Energy Future: Technology and Transformation*. Washington, DC: National Academies Press, 2010.

National Research Council. *Real Prospects for Energy Efficiency in the United States*. Washington, DC: National Academies Press, 2010.

Network for New Energy Choices. *Freeing the Grid*. New York: NNEC, 2010.

Neubauer, Max, et al. "Ka-BOOM! The Power of Appliance Standards." American Council for an Energy-Efficient Economy (ACEEE), Report Number ASAP-7/ACEEE-A091. Washington, DC: ACEEE, 2009.

Newell, Richard. "Shale Gas, A Game Changer for U.S. and Global Gas Markets?" Presented at the Flame—European Gas Conference, Amsterdam, Netherlands March 2, 2010.

Newell, R. G., A. B. Jaffe, and R. N. Stavins. "The Effects of Economic and Policy Incentives on Carbon Mitigation Technologies." *Energy Economics* 28 (2006): 563–578.

Newell, Richard G., Adam B. Jaffe, and Robert N. Stavins. "The Induced Innovation Hypothesis and Energy-Saving Technological Change." *Quarterly Journal of Economics* 114 (3) (August 1999): 941–975.

Nordhaus, William. *A Question of Balance: Weighing the Options on Global Warming Policies*. New Haven: Yale University Press, 2008.

Nuclear Energy Institute. *Credit Subsidy Costs for New Nuclear Power Projects Receiving Department of Energy (DOE) Loan Guarantees: An Analysis of DOE's Methodology and Major Assumptions*. Washington, DC: NEI, 2010.

Office of the Deputy Undersecretary of Defense. *Annual Energy Management Report—Fiscal Year 2009*. Available at http://www.acq.osd.mil/ie/energy/library/aemr_fy_09_may_2010.pdf

Parry, M. L., et al., eds. *Contribution of Working Group II to the Fourth Assessment Report of the Intergovernmental Panel on Climate Change, 2007*. Cambridge: Cambridge University Press, 2007.

Passel, J.S., and D'Vera Cohn. *U.S. Population Projections: 2005–2050*. Washington, DC: Pew Research Center, 2008.

Peretz, Neil. "Standards in Information Technology and the Smart Grid—Are Historical Analogies Useful?" Unpublished MIT IPC Working Paper, May 2010.

Podesta, John, and Karen Kornbluh. "The Green Bank." Center for American Progress, May 21, 2009. Available at http://www.americanprogress.org/issues/2009/05/green_bank.html.

President's Council of Advisors on Science and Technology (PCAST). *Report to the President on Accelerating the Pace of Change in Energy Technologies Through an Integrated Federal Energy Policy*. Washington, DC: Executive Office of the President, November 2010.

Ricker, Amanda. "City to Lay Off Code Enforcement Officer." *Bozeman Daily Chronicle*, June 2, 2011.

Robertson Dan, E., et al. "A New Dawn for Industrial Photosynthesis." *Photosynthesis Research* 13 (February 2011). Available at http://www.springerlink.com/content/j1414q2u5w25h788/fulltext.html.

Romer, Paul. "Implementing a National Technology Strategy with Self-Organizing Industry Investment Boards." *Brookings Papers: Microeconomics* 2 (1993): 345–390.

Rosenberg, Mitchell, and Lynn Hoefgen. "Market Effects and Market Transformation: Their Role in Energy Efficiency Program Design and Evaluation." California Institute for Energy and Environment, March 2009.

Rosenfeld, Arthur H., with Deborah Poskanzer. "A Graph Is Worth a Thousand Gigawatt-Hours: How California Came to Lead the United States in Energy Efficiency." *Innovations* 4 (4) (Fall 2009): 57–80.

Sakhuja, Rohit. "Energy Efficiency in Buildings." Unpublished MIT IPC Working Paper, March 2010.

Sanyal, Paroma, and Linda Cohen. "Powering Progress: Restructuring, Competition, and R&D in the U.S. Electric Utility Industry." *Energy Journal* 30 (2) (2009): 41–80.

Searchinger, T., et al. "Use of U.S. Croplands for Biofuels Increases Greenhouse Gases Through Emissions from Land-Use Change." *Science* 319 ( February 29, 2008): 1238–1240.

Selkowitz, Steve. "Zero Energy Buildings." Presentation at MIT, November 18, 2008.

Shankle, D. L., J. A. Merrick, and T. L. Gilbride. *A History of the Building Energy Standards Program*. Report PNL-9386. Richland, WA: Pacific Northwest Laboratory, 1994.

Smart Grid Legal News. "Maryland rejects Baltimore Gas and Electric's Smart Grid Plan." Available at http://www.smartgridlegalnews.com/regulatory-concerns-1/maryland-will-not-be-the-betamax-of-the-smart-grid/.

Smith, Aaron. *Home Broadband 2010*. Washington, DC: Pew Internet and American Life Project, August 2010.

Sokolov, A. P., et al. "Probabilistic Forecast for Twenty-First-Century Climate Based on Uncertainties in Emissions Without Policy and Climate Parameters." *Journal of Climate* 22 (2009): 5175–5204.

U.S. Department of Energy, Building Energy Codes Program. "Energy Efficiency Contracts Awarded to 24 States." Available at http://www.energycodes.gov/arra/state_funding_awards.stm.

U.S. Department of Energy, Office of Energy Efficiency and Renewable Energy. *Database of State Initiatives for Renewables and Efficiency*. Available at http://www.dsireusa.org.

U.S. Department of Energy. *Buildings Energy Data Book 2009*. Washington, DC: U.S. DOE, 2009.

U.S. Department of Energy. *FY2010 Control Table by Organization*. Available at http://www.cfo.doe.gov/budget/10budget/Content/OrgControl.pdf

U.S. Energy Information Administration. *Electric Power Annual*. Available at http://www.eia.gov/cneaf/electricity/epa/epa_sum.html.

U.S. Energy Information Administration, Office of Coal, Nuclear, Electric and Alternative Fuels. *Federal Financial Interventions and Subsidies in Energy Markets 2007*. Report number SR/CNEAF/2008-01. Washington, DC: E.I.A., 2008.

U.S. Energy Information Administration. *State Energy Data System 2008, Consumption, Price and Expenditure Estimates*. Available at http://www.eia.gov/state/seds/.

U.S. Energy Information Administration. "Emissions of Greenhouse Gases in the U.S." Available at http://www.eia.gov/environment/emissions/ghg_report/ghg_overview.cfm.

U.S. Energy Information Administration. *Alternatives to Traditional Transportation Fuels*. Washington, DC: EIA, 2011.

U.S. Energy Information Administration. *Annual Energy Outlook 2011*. Washington, DC: EIA, 2011.

U.S. Energy Information Administration. *World Shale Gas Resources: An Initial Assessment of 14 Regions Outside the United States*. Washington, DC: U.S. EIA, 2011.

U.S. Energy Information Administration. *The Financial Reporting System*. Available at http://www.eia.gov/emeu/finance/page1a.html.

U.S. Environmental Protection Agency. *Report to Congress on Server and Data Center Energy Efficiency Opportunities, 2007*. Washington, DC: EPA, 2007.

U.S. Federal Communications Commission. *National Broadband Plan*. Washington, DC: FCC, 2010.

U.S. Government Accountability Office. *Agencies Are Taking Steps to Meet High-Performance Federal Building Requirements, but Face Challenges*." Report GAO-10-22. Washington, DC: GAO, 2009.

U.S. Government Accountability Office. *Long-standing Problems with DOE's Program for Setting Efficiency Standards Continue to Result in Forgone Energy Savings*. Report 07-42, January 2007.

U.S. National Aeronautics and Space Administration. *Climate Change: How Do We Know*. Available at http://climate.nasa.gov/evidence.

U.S. National Oceanic and Atmospheric Administration. *State of the Climate: Global Analysis—Annual 2010*. Available at http://www.ncdc.noaa.gov/sotc/global/2010/13.

U.S. National Science Board. *Building a Sustainable Energy Future: U.S. Actions for an Effective Energy Economy Transformation. NSB-09-55*. Washington, DC: NSB, 2009.

U.S. National Science Foundation. *Science and Engineering Indicators 2010*. Washington, DC: NSF, 2010.

U.S. National Science Foundation. *Workshop on the Future Power Engineering Workforce (held November 29–30, 2007), Final Report*. September 5, 2008. Available at http://ecpe.ece.iastate.edu/nsfws/Report%20of%20NSF%20Workshop.pdf.

U.S. National Science Foundation. *Federal R&D Funding by Budget Function: Fiscal Years 2006–08 Detailed Statistical Tables. NSF 07-332*. Washington, DC: NSF, 2007.

U.S. Navy, Energy, Environment, and Climate Change. "Energy." Available at http://greenfleet.dodlive.mil/home/energy.

Willrich, Mason. "Electricity Transmission Policy for America: Enabling a Smart Grid, End-to-End." MIT-IPC-Energy Innovation Working Paper 09-003, July 2009.

Willrich, Mason, and Richard K. Lester. *Radioactive Waste: Management and Regulation*. New York: Free Press, 1977.

Wright, Brian, and Tiffany Shih. Agricultural Innovation. In *Accelerating Energy Innovation: Lessons from Multiple Sectors*, ed. Rebecca M. Henderson and Richard G. Newell. Chicago: University of Chicago Press, 2011.

Zachary, G. Pascal. *Endless Frontier: Vannevar Bush, Engineer of the American Century*. New York: Free Press, 1997.

Zeller, Tom, Jr. "Can We Build in a Brighter Shade of Green?" *New York Times*, September 25, 2010.

Zelnio, Ryan. "The Global Structure of International Scientific Collaborations." Paper presented to the 13th International Society of Scientometrics and Informetrics Conference, Durban, South Africa, July 2011.

Zweibel, Ken, James Mason, and Vasilis Fthenakis. "A Solar Grand Plan." *Scientific American*, January 2008.

# Index